RENEWALS 458-4574

DATE DUE

NOV 14

Broadband Powerline Communications Networks

Broadband Powerline Communications Networks

Network Design

Halid Hrasnica
Abdelfatteh Haidine
Ralf Lehnert

All of
Dresden University of Technology, Germany

John Wiley & Sons, Ltd

Copyright © 2004   John Wiley & Sons Ltd, The Atrium, Southern Gate, Chichester,
West Sussex PO19 8SQ, England

Telephone (+44) 1243 779777

Email (for orders and customer service enquiries): cs-books@wiley.co.uk
Visit our Home Page on www.wileyeurope.com or www.wiley.com

All Rights Reserved. No part of this publication may be reproduced, stored in a retrieval system or transmitted in any form or by any means, electronic, mechanical, photocopying, recording, scanning or otherwise, except under the terms of the Copyright, Designs and Patents Act 1988 or under the terms of a licence issued by the Copyright Licensing Agency Ltd, 90 Tottenham Court Road, London W1T 4LP, UK, without the permission in writing of the Publisher. Requests to the Publisher should be addressed to the Permissions Department, John Wiley & Sons Ltd, The Atrium, Southern Gate, Chichester, West Sussex PO19 8SQ, England, or emailed to permreq@wiley.co.uk, or faxed to (+44) 1243 770620.

This publication is designed to provide accurate and authoritative information in regard to the subject matter covered. It is sold on the understanding that the Publisher is not engaged in rendering professional services. If professional advice or other expert assistance is required, the services of a competent professional should be sought.

Other Wiley Editorial Offices

John Wiley & Sons Inc., 111 River Street, Hoboken, NJ 07030, USA

Jossey-Bass, 989 Market Street, San Francisco, CA 94103-1741, USA

Wiley-VCH Verlag GmbH, Boschstr. 12, D-69469 Weinheim, Germany

John Wiley & Sons Australia Ltd, 33 Park Road, Milton, Queensland 4064, Australia

John Wiley & Sons (Asia) Pte Ltd, 2 Clementi Loop #02-01, Jin Xing Distripark, Singapore 129809

John Wiley & Sons Canada Ltd, 22 Worcester Road, Etobicoke, Ontario, Canada M9W 1L1

Wiley also publishes its books in a variety of electronic formats. Some content that appears in print may not be available in electronic books.

British Library Cataloguing in Publication Data

A catalogue record for this book is available from the British Library

ISBN 0-470-85741-2

Typeset in 10/12pt Times by Laserwords Private Limited, Chennai, India
Printed and bound in Great Britain by TJ International, Padstow, Cornwall
This book is printed on acid-free paper responsibly manufactured from sustainable forestry in which at least two trees are planted for each one used for paper production.

To my parents, with love and respect

H. Hrasnica

Contents

Preface   xi

1 Introduction   1

2 PLC in the Telecommunications Access Area   7
    2.1 Access Technologies   7
       2.1.1 Importance of the Telecommunications Access Area   7
       2.1.2 Building of New Access Networks   8
       2.1.3 Usage of the Existing Infrastructure in the Access Area   11
    2.2 Powerline Communications Systems   14
       2.2.1 Historical Overview   14
       2.2.2 Power Supply Networks   14
       2.2.3 Standards   15
       2.2.4 Narrowband PLC   16
       2.2.5 Broadband PLC   19
    2.3 PLC Access Networks   19
       2.3.1 Structure of PLC Access Networks   19
       2.3.2 In-home PLC Networks   21
       2.3.3 PLC Network Elements   22
       2.3.4 Connection to the Core Network   27
       2.3.5 Medium-voltage PLC   31
    2.4 Specific PLC Performance Problems   32
       2.4.1 Features of PLC Transmission Channel   33
       2.4.2 Electromagnetic Compatibility   33
       2.4.3 Impact of Disturbances and Data Rate Limitation   34
       2.4.4 Realization of Broadband PLC Transmission Systems   36
       2.4.5 Performance Improvement by Efficient MAC Layer   36
    2.5 Summary   37

3 PLC Network Characteristics   39
    3.1 Network Topology   39
       3.1.1 Topology of the Low-voltage Supply Networks   39

		3.1.2	Organization of PLC Access Networks	41

		3.1.2	Organization of PLC Access Networks	41
		3.1.3	Structure of In-home PLC Networks	47
		3.1.4	Complex PLC Access Networks	48
		3.1.5	Logical Network Models	50
	3.2	Features of PLC Transmission Channel		52
		3.2.1	Channel Characterization	52
		3.2.2	Characteristics of PLC Transmission Cable	53
		3.2.3	Modeling of the PLC Channel	54
	3.3	Electromagnetic Compatibility of PLC Systems		55
		3.3.1	Different Aspects of the EMC	56
		3.3.2	PLC EM Disturbances Modeling	61
		3.3.3	EMC Standards for PLC Systems	65
	3.4	Disturbance Characterization		70
		3.4.1	Noise Description	70
		3.4.2	Generalized Background Noise	71
		3.4.3	Impulsive Noise	73
		3.4.4	Disturbance Modeling	74
	3.5	Summary		76
4	**Realization of PLC Access Systems**			**79**
	4.1	Architecture of the PLC Systems		79
	4.2	Modulation Techniques for PLC Systems		82
		4.2.1	Orthogonal Frequency Division Multiplexing	82
		4.2.2	Spread-Spectrum Modulation	89
		4.2.3	Choice of Modulation Scheme for PLC Systems	95
	4.3	Error Handling		97
		4.3.1	Overview	97
		4.3.2	Forward Error Correction	98
		4.3.3	Interleaving	108
		4.3.4	ARQ Mechanisms	111
	4.4	PLC Services		114
		4.4.1	PLC Bearer Service	114
		4.4.2	Telecommunications Services in PLC Access Networks	115
		4.4.3	Service Classification	121
	4.5	Summary		123
5	**PLC MAC Layer**			**125**
	5.1	Structure of the MAC Layer		125
		5.1.1	MAC Layer Components	125
		5.1.2	Characteristics of PLC MAC Layer	126
		5.1.3	Requirements on the PLC MAC Layer	126
	5.2	Multiple Access Scheme		128
		5.2.1	TDMA	129
		5.2.2	FDMA	132
		5.2.3	CDMA	135
		5.2.4	Logical Channel Model	150

	5.3	Resource-sharing Strategies	151
		5.3.1 Classification of MAC Protocols	153
		5.3.2 Contention Protocols	154
		5.3.3 Arbitration Protocols	169
		5.3.4 IEEE 802.11 MAC Protocol	177
	5.4	Traffic Control	181
		5.4.1 Duplex Mode	181
		5.4.2 Traffic Scheduling	185
		5.4.3 CAC Mechanism	189
	5.5	Summary	192
6	**Performance Evaluation of Reservation MAC Protocols**		**195**
	6.1	Reservation MAC Protocols for PLC	195
		6.1.1 Reservation Domain	196
		6.1.2 Signaling Procedure	198
		6.1.3 Access Control	199
		6.1.4 Signaling MAC Protocols	203
	6.2	Modeling PLC MAC Layer	205
		6.2.1 Analysis Method	205
		6.2.2 Simulation Model for PLC MAC Layer	208
		6.2.3 Traffic Modeling	211
		6.2.4 Simulation Technique	215
	6.3	Investigation of Signaling MAC Protocols	218
		6.3.1 Basic Protocols	218
		6.3.2 Protocol Extensions	230
		6.3.3 Advanced Polling-based Reservation Protocols	236
	6.4	Error Handling in Reservation MAC Protocols	244
		6.4.1 Protection of the Signaling Procedure	244
		6.4.2 Integration of ARQ in Reservation MAC Protocols	246
		6.4.3 ARQ for Per-packet Reservation Protocols	247
	6.5	Protocol Comparison	250
		6.5.1 Specification of Required Slot Structure	250
		6.5.2 Specification of Traffic Mix	252
		6.5.3 Simulation Results	253
		6.5.4 Provision of QoS in Two-step Reservation Protocol	254
	6.6	Summary	256
Appendix A			**259**
	A.1	Abbreviations	259
References			**263**
Index			**273**

Preface

Access networks implement the inter-connection of the customers/subscribers to wide-area communication networks. They allow a large number of subscribers to use various telecommunications services. However, the costs of realization, installation, and maintenance of access networks are very high, very often representing more than 50% of the investment in the network. Therefore, network providers try to realize the access network at as low a cost as possible to increase their competitiveness in the deregulated telecommunications market. In most cases, access networks are still the property of incumbent network providers (e.g., the former monopolistic telephone companies). Because of that, new network providers try to find solutions to realize their own access networks. A promising possibility for the realization of access networks is offered by the PowerLine Communications (PLC) technology.

PowerLine Communications technology allows the usage of electrical power supply networks for communications purposes and, today, also broadband communication services. The main idea behind PLC is the reduction in operational costs and expenditure for realization of new telecommunications networks. Using electrical supply networks for telecommunications has also been known since the beginning of the twentieth century. Thus high-, medium- and low-voltage supply networks have been used for internal communications of electrical utilities and for the realization of remote measuring and control tasks. PLC is also used in internal electrical installations within buildings and homes (the so-called in-home PLC) for various communications applications. Generally, we can divide PLC systems into two groups: narrowband PLC allowing communications services with relatively low data rates (up to 100 kbps) and ensuring realization of various automation and control applications as well as a few voice channels, and broadband PLC systems allowing data rates beyond 2 Mbps and, accordingly, realization of a number of typical telecommunications services in parallel, such as telephony and internet access.

Broadband PLC in low-voltage supply networks seems to be a cost-effective solution for "last mile" communications networks, the so-called PLC access networks. Nowadays, there are many activities concerned with the development and application of PLC technology in the access area. Thus, we find a number of manufacturers offering PLC products that ensure data rates between 2 and 4 Mbps and announcing new PLC systems with data rates up to 45 Mbps or more. There are also numerous PLC field trials worldwide, as well as several PLC access networks in commercial use. The number of PLC subscribers is still growing. A similar development in medium-voltage and in-home PLC networks is in progress as well. On the other hand, there are no existing standards for broadband PLC networks, which are supposed to use a frequency range up to 30 MHz. In particular, the problem of electromagnetic compatibility of PLC systems with reference to their coexistence with other telecommunications systems, such as various radio services, has not yet been completely solved. Therefore, PLC technology is now in a very important development phase that will determine its future, its application areas, and its penetration into telecommunications world in competition with other broadband technologies.

Because of the absence of standards and, understandable, detailed publication of sensitive research material by PLC manufacturers, there is very little information on broadband PLC systems and networks in the literature. We find a number of papers, several dissertations, and a few books covering different, mainly very specific, research areas, which are not suitable for the wider community of readers. On the other hand, there are many publications describing general PLC-related topics but without, or with very little, technical content. Therefore, it is necessary to provide complete information on broadband PLC networks that includes both general information on PLC technology and also offers technical details that are important for the realization of PLC systems. The book "Powerline Communications" by Klaus Dostert covers mainly narrowband PLC technology, and it focuses more on the transmission aspects of PLC.

This book contributes to the design aspects of broadband PLC access systems and their network components. The intention of this book is to explain how broadband PLC networks are realized; what the important characteristics, as well as environment, for the transmission through electrical power grids are; and what implementation solutions have been considered recently for the realization of broadband PLC systems.

The authors of this book, all of them from the Chair for Telecommunications at Dresden University of Technology – Germany, have been involved in the research and development of PLC networks and systems for several years. Our department has participated in several international industry and EU supported research projects on PLC and cooperated with a number of partners also involved in the actual development of this technology. The chair is a member of the PLC Forum. The authors have published more than 20 research publications on broadband PLC access networks, performance evaluation of PLC systems, modeling PLC networks, and development of PLC MAC layer and its protocols. In our department, we have developed a simulation tool called PAN-SIM (PLC Access Network Simulator), used for performance analysis of PLC networks, which has also been presented in several trade fairs and specialized conferences.

This book has been written for the following groups of readers:

- Lecturers (professors, PhD researchers), for research and educational purposes at universities
- Developers of PLC equipment, systems, interfaces, and so on.
- Network engineers at potential PLC network operators
- Business people, managers, or policy makers who need an overview of PLC technology and its possibilities, and of course
- Students with an interest in PLC and other telecommunications technologies.

During our work on this book, many people have supported us in different ways. Therefore, we would like to thank them. First, we would like to thank all our colleagues at the Chair for Telecommunications, Dresden University of Technology – Germany, for their valuable professional help and for creating the friendly atmosphere in our department that really helped us complete this project. We also have drawn considerably from our involvement in several research projects. Therefore, we would like to thank all our partners in the PALAS project in the 5th framework programme of the European Community and our colleagues from Regiocom (Magdeburg) and Drewag (Dresden). Our sincere thanks go to all the students who helped us during the work on PLC and to numerous colleagues worldwide with whom we had very useful discussions on various occasions.

Dresden, January 2004

Halid Hrasnica, Abdelfatteh Haidine, Ralf Lehnert

1

Introduction

During the last decades, the usage of telecommunications systems has increased rapidly. Because of a permanent necessity for new telecommunications services and additional transmission capacities, there is also a need for the development of new telecommunications networks and transmission technologies. From the economic point of view, telecommunications promise big revenues, motivating large investments in this area. Therefore, there are a large number of communications enterprises that are building up high-speed networks, ensuring the realization of various telecommunications services that can be used worldwide. However, the investments are mainly provided for transport networks that connect various communications nodes of different network providers, but do not reach the end customers. The connection of the end customers to a transport network, as part of a global communications system, is realized over distribution and access networks (Fig. 1.1). The distribution networks cover larger geographical areas and realize connection between access and transport networks, whereas the access networks cover relatively smaller areas.

The direct connection of the customers/subscribers is realized over the access networks, realizing access of a number of subscribers situated within a radius of several hundreds of meters. However, the costs for realization, installation and maintenance of the access networks are very high. It is usually calculated that about 50% of all network investments belongs to the access area. On the other hand, a longer time is needed for paying back the invested capital because of the relatively high costs of the access networks, calculated per connected subscriber. Therefore, the network providers try to realize the access network with possibly low costs.

After the deregulation of the telecommunications market in a large number of countries, the access networks are still the property of incumbent network providers (former monopolistic telephone companies). Because of this, the new network providers try to find a solution to offer their own access network. An alternative solution for the realization of the access networks is offered by the PLC (PowerLine Communications) technology using the power supply grids for communications. Thus, for the realization of the PLC networks, there is no need for the laying of new communications cables. Therefore, application of PLC in low-voltage supply networks seems to be a cost-effective solution for so-called "last mile" communications networks, belonging to the access area. Nowadays, network subscribers use various telecommunications services with higher data rates and

---

Broadband Powerline Communications Networks   H. Hrasnica, A. Haidine, and R. Lehnert
© 2004 John Wiley & Sons, Ltd   ISBN: 0-470-85741-2

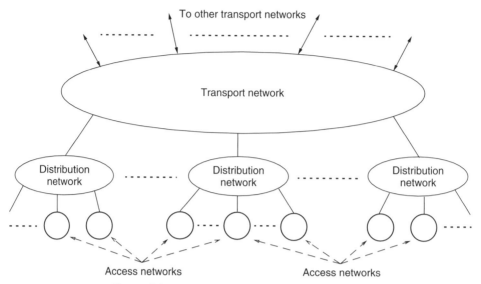

Figure 1.1 Telecommunications network hierarchy

QoS (Quality of Service) requirements. PLC systems applied in the access area that ensure realization of telecommunications services with the higher QoS requirements are called "broadband PLC access networks". The contribution of this book is directed to give a set of information that is necessary to be considered for the design of the broadband PLC access systems and their network components.

To make communications in a power supply network possible, it is necessary to install so-called *PLC modems*, which ensure transmission of data signals over the power grids (Fig. 1.2). A PLC modem converts a data signal received from conventional communications devices, such as computers, telephones, and so on, in a form that is suitable for transmission over powerlines. In the other transmission direction, the modem receives a

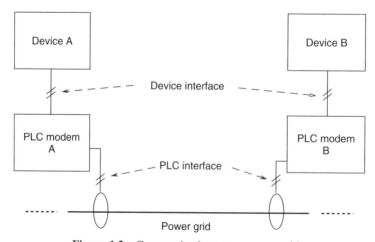

Figure 1.2 Communications over power grids

data signal from the power grids and after conversion delivers it to the communications devices. Thus, the PLC modems, representing PLC-specific communications equipment, provide a necessary interface for interconnection of various communications devices over power supply networks. The PLC-specific communications devices, such as PLC modems, have to be designed to ensure an efficient network operation under transmission conditions, typical for power supply networks and their environment.

However, power supply networks are not designed for communications and they do not present a favorable transmission medium. Thus, the PLC transmission channel is characterized by a large, and frequency-dependent attenuation, changing impedance and fading as well as unfavorable noise conditions. Various noise sources, acting from the supply network, due to different electric devices connected to the network, and from the network environment, can negatively influence a PLC system, causing disturbances in an error-free data transmission. On the other hand, to provide higher data rates, PLC networks have to operate in a frequency spectrum of up to 30 MHz, which is also used by various radio services. Unfortunately, a PLC network acts as an antenna producing electromagnetic radiation in its environment and disturbs other services operating in the same frequency range. Therefore, the regulatory bodies specify very strong limits regarding the electromagnetic emission from the PLC networks, with a consequence that PLC networks have to operate with a limited signal power. This causes a reduction of network distances and data rates and increases sensitivity to disturbances.

The reduction of the data rates is particularly disadvantageous because of the fact that PLC access networks operate in a shared transmission medium, in which a number of subscribers compete to use the same transmission resources (Fig. 1.3). In the case of PLC

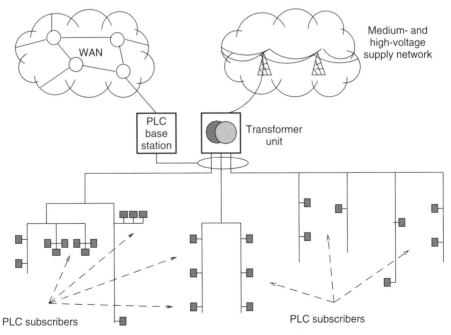

Figure 1.3 PLC access network

access networks, the transmission medium provided by a low-voltage supply network is used for communication between the subscribers and a so-called *PLC base station*, which connects the access network to a wide area network (WAN) realized by conventional communications technology.

To reduce the negative impact of powerline transmission medium, PLC systems have to apply efficient modulation, such as spread spectrum and Orthogonal Frequency Division Multiplexing (OFDM). The problem of disturbances can also be solved by well-known error-handling mechanisms (e.g. forward error correction (FEC), Automatic Repeat reQuest (ARQ)). However, their application consumes a certain portion of the PLC network capacity because of overhead and retransmission. On the other hand, a PLC access network has to be economically efficient, serving possibly a large number of subscribers. This can be ensured only by a good utilization of the limited network capacity. Simultaneously, PLC systems have to compete with other access technologies (e.g. digital subscriber line (DSL), cable television (CATV)) and to offer different telecommunications services with a satisfactory QoS. Both good network utilization and provision of QoS guarantees can be achieved by an efficient Medium Access Control (MAC) layer.

Nowadays, there are no existing standards or specifications considering physical and MAC layers for PLC access networks. The manufacturers of the PLC equipment developed proprietary solutions for the MAC layer that are incompatible with each other. Therefore, we consider various solutions for realization of both physical and MAC layers in broadband PLC access networks to be implemented in PLC-specific communications equipment, such as PLC modems (Fig. 1.3). Detailed description of the PLC physical layer, including consideration of the PLC network characteristics, such as transmission features and noise behavior, and consideration of modulation schemes for PLC, can also be found in another available book on this topic, "Powerline Communications", written by Prof. Dostert [Dost01], in which both the narrowband and broadband PLC systems are considered. In this book, we focus on the broadband access networks and describe characteristics of the physical layer and applied modulation schemes for the broadband PLC systems, and introduce an investigation of PLC MAC layer. Nowadays, the issue of the PLC MAC layer is only considered in a few scientific publications (e.g. [Hras03]). Therefore, in this book we emphasize a consideration of the MAC layer and its protocols to be applied in the broadband PLC access networks.

The book is organized as follows: in Chapter 2, we discuss the role of PLC in telecommunications access area and present basics about narrowband and broadband PLC systems, network structure with its elements and PLC-specific performance problems that have to be overcome for realization of broadband access networks. The characteristics of the PLC transmission medium are presented in Chapter 3, which includes a topology analysis of the low-voltage supply networks, description of the electromagnetic compatibility issue (EMC) in broadband PLC, noise characterization and disturbance modeling, as well as a description of the PLC transmission channel and its features. In Chapter 4, we present a protocol architecture for PLC networks and define PLC-specific network layers. Later, we describe spread spectrum and OFDM modulation schemes, which are outlined as favorable solutions for PLC. Furthermore, various possibilities for realization of error handling in PLC systems are considered. Finally, in Chapter 4, we analyze telecommunications services to be used in PLC networks and specify traffic models for their representation in investigations of the PLC networks. The MAC layer, as a part of the common PLC protocol architecture, is separately analyzed in Chapter 5. We introduce different solutions of multiple-access schemes and consider various MAC protocols for their application in PLC. Furthermore, several solutions for traffic control in PLC networks are discussed. Finally, in Chapter 6, we present a comprehensive performance evaluation of reservation MAC protocols, which are outlined as a suitable solution for application in broadband PLC access networks. In this investigation, we compare various signaling MAC protocols under different traffic and disturbance conditions, representing a typical user and noise behavior expected in broadband PLC access networks.

2

PLC in the Telecommunications Access Area

2.1 Access Technologies

2.1.1 Importance of the Telecommunications Access Area

Access networks are very important for network providers because of their high costs and the possibility of the realization of a direct access to the end users/subscribers. Lately, about 50% of all investments in the telecommunications infrastructure is needed for the realization of telecommunications access networks. However, an access network connects a limited number of individual subscribers, as opposed to a transport communication network (Fig. 2.1). Therefore, economic efficiency of the access networks is significantly lower than in wide area networks (WAN).

In the case of the so-called *big customers* (business, governmental or industrial customers), the access networks connect a higher number of subscribers who are concentrated within a building or in a small region (e.g. campus). The big customers usually use various telecommunication services intensively and bring high sales to the network providers. Therefore, the realization of particular access networks for the big customers makes economical sense.

As opposed to the big customers, individual subscribers (e.g. private subscribers, Fig. 2.1) use the telecommunication services less intensively. Accordingly, realization of the access networks for individual subscribers is also economically less efficient. On the other hand, a direct access to the subscribers increases the opportunities for network providers to offer a higher number of various services. This attracts the subscribers to become contract-bound customers of a particular network provider, which increases the usage of its transport network. Therefore, the access to the individual subscribers seems to be important as well.

After the deregulation of the telecommunications market in a large number of countries, the access networks are still the property of former monopolistic companies (incumbent network providers). New network providers build up their transport networks (WAN), but they still have to use the access infrastructure owned by an incumbent provider. Because of this, new network providers try to find a solution to offer their own access network

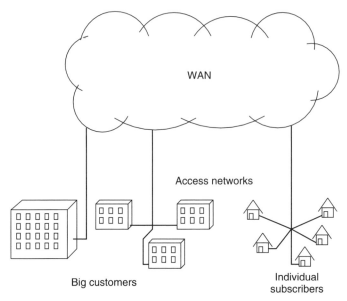

Figure 2.1 General structure of telecommunications networks

to the subscribers. On the other hand, a rapid development of new telecommunications services increases the demand for more transmission capacity in the transport networks as well as in the access area. Therefore, there is a permanent need for an extension of the access infrastructure. There are two possibilities for the expansion of the access networks:

- Building of new networks or
- Usage of the existing infrastructure.

Building of new access networks is the best way to implement the newest communications technology, which allows realization of very attractive services. On the other hand, building of new access networks is expensive. Thus, the usage of the existing infrastructure for realization of the access networks is a more attractive solution for network providers because of lower costs. However, the existing infrastructure has to be renewed and equipped to be able to offer attractive telecommunications services as well.

2.1.2 Building of New Access Networks

Generally, the building of new access networks can be realized with the following techniques:

- New cable or optical network
- Wireless access systems
- Satellite systems.

Nowadays, the optical telecommunications networks offer higher data rates than any other communications technology. Frequent usage of optical transmission systems within transport networks (WAN) reduces their costs. Therefore, the implementation of optical communications networks also becomes economically efficient in the access area. This allows realization of a sufficient transmission capacity and attractive services.

However, laying of new optical or cable networks is very costly because of the required voluminous construction steps. Very often, it has to be carried out within urban areas causing legal problems and additional costs. Finally, the building of new cable or optical networks takes a long time. Because of these reasons, laying of new networks is mostly done in new settlements and areas with a big subscriber concentration (business and governmental centers, dense industrial areas, etc.).

To avoid realization of new cable or optical networks, various wireless transmission systems can be applied in the access area. The two approaches that can be applied for the realization of wireless access networks can be distinguished as follows [GargSn96]:

- Wireless mobile systems
- Fixed wireless systems.

Well-known wireless mobile systems are DECT, GSM/GPRS, and UMTS. Mobile networks provide a large number of cells to cover a wide communication area, which ensures a permanent connection for mobile subscribers in the area covered (cellular network, Fig. 2.2). A frequency range is allocated to each cell allowing communication between mobile terminals (MT) and base stations. Different frequencies (or codes for UMTS) are allocated to neighbors' cells to avoid interferences between them. Generally, a base station covers a number of wireless communication cells connecting them to a WAN. The wireless mobile systems offer sum transmission data rates up to 2 Mbps.

Fixed wireless systems, called WLL systems (Wireless Local Loop), are more suitable for application in the access area than the mobile systems [GargSn96]. WLL systems also provide base stations that connect a number of subscribers situated in a relatively small area (Fig. 2.3). As opposed to mobile wireless systems, WLL subscribers have a fixed position with antennas that are located on high posts on buildings or houses. Therefore, WLL systems provide constant propagation paths between the base station and

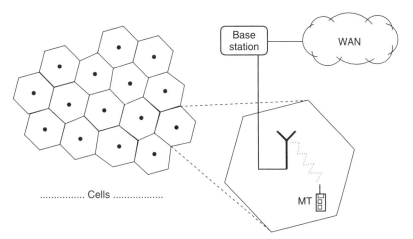

Figure 2.2 Structure of wireless mobile networks

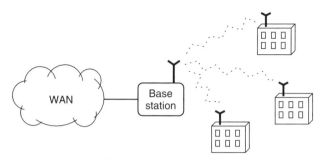

Figure 2.3 Structure of fixed wireless networks – Wireless Local Loop

the subscribers, and, accordingly, provide a better SNR (signal-to-noise ratio) behavior than in the wireless mobile systems. The data rates are also higher than in the mobile systems; up to 10 Mbps in the downlink transmission direction (from the base station to the subscribers) and up to 256 kbps in the uplink (from the subscribers to the base station). However, the data rates realized in different WLL systems are still increasing.

WLL systems realize connections between a base station and the appropriate customer transreceiver station equipped with an antenna. A customer station usually covers a building or a house with a number of individual subscribers using various communications services. The connection between a customer station and its subscribers can be realized in different ways, via a wireline communications infrastructure or as a home wireless network.

Nowadays, the home wireless networks are realized as the so-called *Wireless Local Area Networks* (WLAN). A WLAN operates usually within buildings and covers a relatively small area, ensuring data rates beyond 20 Mbps (see e.g. [Walke99]). WLAN systems are used to cover a number of rooms within business premises or private households (e.g. to cover a house with the belonging surroundings, garden, etc.). For this purpose, one or more antennas are installed, which makes possible the usage of various communications devices in the entire covered area, without a need for any kind of wireline connections. The antennas are situated in the so-called *access points* (AP, Fig. 2.4), which are usually connected to a wireline network. In this way, a WLAN is connected to the network servers and to WAN. Thus, the mobile terminals of a WLAN are able to use various services and access the global communication network.

Both mobile and fixed wireless systems are still expensive for application in access networks. Furthermore, coverage of large areas with wireless systems needs a higher number of base stations and antennas, which, additionally, increases the network costs. Lastly, the maximum data rates reached in WLANs are significantly lower than the data rates in optical networks.

The third possibility for the realization of the access networks are satellite systems, which are nowadays mostly used for worldwide long-distance communications. The low Earth orbit (LEO) and medium Earth orbit (MEO) satellites were developed for application in the communications access area [Dixi99]. Such satellite systems, like the Iridium system [HubbSa97], should extend the existing cellular systems in which the base stations are replaced (or partly replaced) by the satellites. However, the satellite access systems currently do not provide good economic efficiency and some satellite projects have recently been canceled (e.g. Iridium).

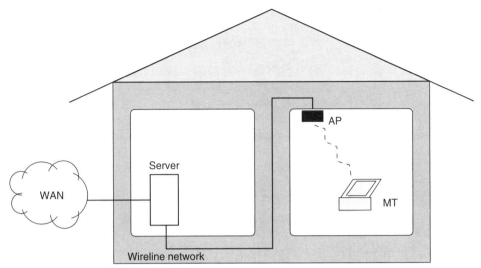

Figure 2.4 Structure of Wireless Local Area Networks – WLAN

2.1.3 Usage of the Existing Infrastructure in the Access Area

The building of expensive new communications networks can be avoided by the usage of the existing infrastructure for the realization of access networks. In this case, already existing wireline networks are used to connect the subscribers to the transport telecommunications networks. The following networks can be used for this purpose:

- Classical telephone networks
- TV cable networks (CATV)
- Electrical power supply networks.

Nowadays, the classic telephone networks are equipped by Digital Subscriber Line (DSL) systems to provide higher data rates in the access area. Asymmetric Digital Subscriber Line (ADSL) is a variant of DSL technology, mostly applied in the access networks (e.g. operated by the Deutsche Telekom – German Telecom) [OrthPo99]. The ADSL technique can ensure up to 8 Mbps in downlink transmission direction and up to 640 kbps in the uplink [Ims99] under optimal conditions (length, transmission features of lines, etc.).

The subscribers using DSL access systems are connected to a central switching node (e.g. local exchange office) over a star formed network, which allows each DSL subscriber to use the full data rates (Fig. 2.5). The central nodes are usually connected to the backbone network (WAN) over a distribution system using high-speed optical transmission technology.

For the realization of DSL access networks, appropriate equipment is needed on the subscriber side (e.g. ADSL modem) as well as within the central node. Generally, ADSL modems on the subscriber side connect various communications devices to the transmission line. Nowadays, the most applied communications service using DSL technique is broadband Internet access. However, there is a possibility of the realization of classical telephone service as well as advanced services providing transmission of various video signals (e.g.

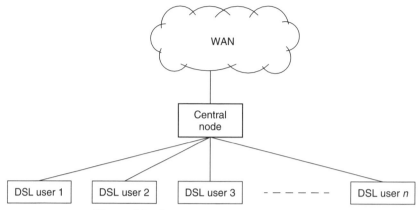

Figure 2.5 Structure of DSL access networks

pay and broadcast TV, interactive video, etc.). The central node provides a number of modems connecting the individual subscribers and acts as a concentrator, so-called DSL access multiplexer, connecting DSL end user to the backbone communications network.

So, for the realization of DSL-based access networks, it is only necessary to install the appropriate modems on both the subscriber and the central node side. However, in some cases there is a need for a partial reconstruction and improvement of the subscribers' lines, if the physical network features do not fulfill requirements for the realization of DSL access. The maximum data rates in DSL systems depend on the length of the subscribers' lines and their transmission characteristics. Table 2.1 presents an overview of different DSL techniques and their features.

Table 2.1 Characteristics of xDSL systems [Dixi99]

Acronym	Name	Data rate	Mode	Max. dist. (km)	Number of wire pairs
DSL	Digital subscriber line	160 kbps	Duplex	6	One
HDSL	High-data-rate DSL	1.544 Mbps 2.048 Mbps	Duplex	4	Two, Three
SDSL	Single-line DSL	1.544 Mbps 2.048 Mbps	Duplex	3	One
ADSL	Asymmetric DSL	1.5 to 6.144 Mbps 16 to 640 kbps	Downlink Uplink	4 to 6	One
RADSL	Rate-adaptive DSL	Adaptive to ADSL rates	Downlink Uplink	4 to 6	One
VDSL	Very-high-data-rate DSL	13 to 52 Mbps 1.5 to 2.3 Mbps	Downlink Uplink	0.3 to 1.5	One
(A)DSL Lite (or UADSL)	ADSL Lite or Universal ADSL	1.5 Mbps 512 kbps	Downlink Uplink	6	One

CATV networks are designed for the broadcasting of TV programs to homes, but they are also very often used for the realization of other telecommunications services. In some regions, CATV networks are widely available and connect a very large number of end users. Also, cabling technique used for CATV wire infrastructure has to ensure higher data rates providing transmission of multiple TV channels with a certain quality. Therefore, CATV networks seem to be an alternative solution for the realization of access networks too. The access systems realized over CATV networks offer up to 50 Mbps in downlink and up to 5 Mbps in uplink transmission direction [Ims99, Hern97]. However, on average there are about 600 subscribers connected to a CATV access network who have to share the common network capacity – shared medium (Fig. 2.6).

The subscribers of a CATV access network are connected to a central node, similar to DSL access networks. The appropriate modems, so-called *cable modems*, are also needed on both the subscriber and the central node side. The subscribers of a CATV system equipped to serve as a general access network are able to use various communication services as well. However, within the network there are amplifiers that usually operate only in the downlink direction, because the original purpose of CATV networks is to transmit TV signals from a central antenna to the subscribers. Therefore, the amplifiers have to be modified to operate in both transmission directions, allowing bidirectional data transmission, which is needed for the realization of access networks.

Telephone networks usually belong to former monopolistic companies (incumbent providers) and this is a big disadvantage for the new network providers to use them to offer services like ADSL. It is also very often the case with the CATV networks. Additionally, the CATV networks have to be made capable of bidirectional transmission, which results in extra costs. In some cases, the subscriber lines have to be modified to ensure application of DSL technology, which increases the cost as well. Because of these reasons, the usage of power supply systems for communication seems to be a reasonable solution for the realization of alternative access networks. However, PowerLine Communications (PLC) technology should provide an economically efficient solution and should offer a big palette of the telecommunications services with a certain quality to be able to compete with other access technologies.

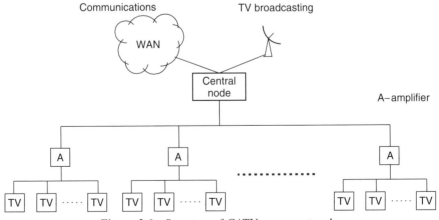

Figure 2.6 Structure of CATV access networks

2.2 Powerline Communications Systems

2.2.1 Historical Overview

PowerLine Communications is the usage of electrical power supply networks for communications purposes. In this case, electrical distribution grids are additionally used as a transmission medium for the transfer of various telecommunications services. The main idea behind PLC is the reduction of cost and expenditure in the realization of new telecommunications networks.

High- or middle-voltage power supply networks could be used to bridge a longer distance to avoid building an extra communications network. Low-voltage supply networks are available worldwide in a very large number of households and can be used for the realization of PLC access networks to overcome the so-called telecommunications "last mile". Powerline communications can also be applied within buildings or houses, where an internal electrical installation is used for the realization of in-home PLC networks.

The application of electrical supply networks in telecommunications has been known since the beginning of the twentieth century. The first Carrier Frequency Systems (CFS) had been operated in high-voltage electrical networks that were able to span distances over 500 km using 10-W signal transmission power [Dost97]. Such systems have been used for internal communications of electrical utilities and realization of remote measuring and control tasks. Also, the communications over medium- and low-voltage electrical networks has been realized. Ripple Carrier Signaling (RCS) systems have been applied to medium- and low-voltage networks for the realization of load management in electrical supply systems.

Internal electrical networks have been mostly used for realization of various automation services. Application of in-home PLC systems makes possible the management of numerous electrical devices within a building or a private house from a central control position without the installation of an extra communications network. Typical PLC-based building automation systems are used for security observance, supervision of heating devices, light control, and so on.

2.2.2 Power Supply Networks

The electrical supply systems consist of three network levels that can be used as a transmission medium for the realization of PLC networks (Fig. 2.7):

- High-voltage (110–380 kV) networks connect the power stations with large supply regions or big customers. They usually span very long distances, allowing power exchange within a continent. High-voltage networks are usually realized with overhead supply cables.
- Medium-voltage (MV) (10–30 kV) networks supply larger areas, cities and big industrial or commercial customers. Spanned distances are significantly shorter than in the high-voltage networks. The medium-voltage networks are realized as both overhead and underground networks.
- Low-voltage (230/400 V, in the USA 110 V) networks supply the end users either as individual customers or as single users of a bigger customer. Their length is usually up to a few hundred meters. In urban areas, low-voltage networks are realized with underground cables, whereas in rural areas they exist usually as overhead networks.

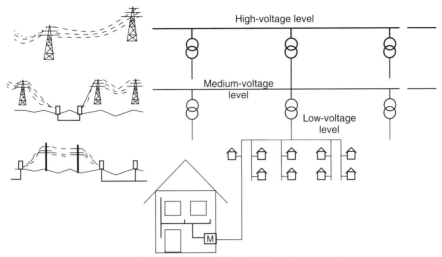

Figure 2.7 Structure of electrical supply networks

In-home electrical installations belong to the low-voltage network level. However, internal installations are usually owned by the users. They are connected to the supply network over a meter unit (M). On the other hand, the rest of the low-voltage network (outdoor) belongs to the electrical supply utilities.

Low-voltage supply networks directly connect the end customers in a very large number of households worldwide. Therefore, the application of PLC technology in low-voltage networks seems to have a perspective regarding the number of connected customers. On the other hand, low-voltage networks cover the last few hundreds of meters between the customers and the transformer unit and offer an alternative solution using PLC technology for the realization of the so-called "last mile" in the telecommunications access area.

2.2.3 Standards

The communications over the electrical power supply networks is specified in a European standard CENELEC EN 50065, providing a frequency spectrum from 9 to 140 kHz for powerline communications (Tab. 2.2). CENELEC norm significantly differs from American and Japanese standards, which specify a frequency range up to 500 kHz for the application of PLC services.

Table 2.2 CENELEC bands for powerline communications

Band	Frequency range (kHz)	Max. transmission amplitude (V)	User dedication
A	9–95	10	Utilities
B	95–125	1.2	Home
C	125–140	1.2	Home

CENELEC norm makes possible data rates up to several thousand bits per second, which are sufficient only for some metering functions (load management for an electrical network, remote meter reading, etc.), data transmission with very low bit rates and the realization of few numbers of transmission channels for voice connections. However, for application in modern telecommunications networks, PLC systems have to provide much higher data rates (beyond 2 Mbps). Only in this case, PLC networks are able to compete with other communications technologies, especially in the access area (Sec. 2.1).

For the realization of the higher data rates, PLC transmission systems have to operate in a wider frequency spectrum (up to 30 MHz). However, there are no PLC standards that specify the operation of PLC systems out of the frequency bands defined by the CENELEC norm. Currently, there are several bodies that try to lead the way for standardization of broadband PLC networks, such as the following:

- PLCforum [PLCforum] is an international organization with the aim to unify and represent the interests of players engaged in PLC from all over the world. There are more than 50 members in the PLCforum; manufacturer companies, electrical supply utilities, network providers, research organizations, and so on. PLCforum is organized into four working groups: Technology, Regulatory, Marketing and Inhouse working group.
- The HomePlug Powerline Alliance [HomePlug] is a not-for-profit corporation formed to provide a forum for the creation of open specifications for high-speed home powerline networking products and services. HomePlug is concentrated on in-home PLC solutions and it works close to PLCforum as well.

Standardization activities for broadband PLC technology are also included in the work of European Telecommunications Standards Institute (ETSI) and CENELEC.

2.2.4 Narrowband PLC

The narrowband PLC networks operate within the frequency range specified by the CENELEC norm (Tab. 2.2). This frequency range is divided into three bands: A, to be used by power supply utilities, and B and C, which are provided for private usage. The utilities use narrowband PLC for the realization of the so-called *energy-related services*. Frequency bands B and C are mainly used for the realization of building and home automation. Nowadays, the narrowband PLC systems provide data rates up to a few thousand bits per second (bps) [Dost01]. The maximum distance between two PLC modems can be up to 1 km. To overcome longer distances, it is necessary to apply a repeater technique.

The narrowband PLC systems apply both narrowband and broadband modulation schemes. First narrowband PLC networks have been realized by the usage of Amplitude Shift Keying (ASK) [Dost01]. The ASK is not robust against disturbances and, therefore, is not suitable for application in PLC networks. On the other hand, Binary Phase Shift Keying (BPSK) is a robust scheme and, therefore, is more suitable for application in PLC. However, phase detection, which is necessary for the realization of BPSK, seems to be complex and BPSK-based systems are not commonly used. Most recent narrowband PLC systems apply Frequency Shift Keying (FSK), and it is expected that BPSK will be used in future communications systems [Dost01].

Broadband modulation schemes are also used in narrowband PLC systems. The advantages of broadband modulation, such as various variants of spread spectrum, are its robustness against narrowband noise and the selective attenuation effect that exists in the PLC networks [Dost01]. A further transmission scheme also used in narrowband PLC system is Orthogonal Frequency Division Multiplexing (OFDM) [Bumi03].

A comprehensive description of various narrowband PLC systems, including their realization and development, can be found in [Dost01]. The aim of this book is a presentation of broadband PLC systems, and, therefore, the narrowband systems are not discussed in detail. However, to sketch the possibilities of the narrowband PLC, we present several examples for application of this technology in the description below.

A very important area for the application of narrowband PLC is building/home automation. PLC-based automation systems are realized without the installation of additional communications networks (Fig. 2.8). Thus, the high costs that are necessary for the installation of new networks within existing buildings can be significantly decreased by the usage of PLC technology. Automation systems realized by PLC can be applied to different tasks to be carried out within buildings:

- Control of various devices that are connected to the internal electro installation, such as illumination, heating, air-conditioning, elevators, and so on.
- Centralized control of various building systems, such as window technique (darkening) and door control.
- Security tasks; observance, sensor interconnection, and so on.

PLC-based automation systems are not only used in large buildings but they are also very often present in private households for the realization of similar automation tasks (home automation). In this case, several authors talk about so-called *smart homes*.

A PLC variant of the EIB (European Installation BUS) standard is named *Powernet-EIB*. PLC modems designed according to the Powernet-EIB can be easily connected to

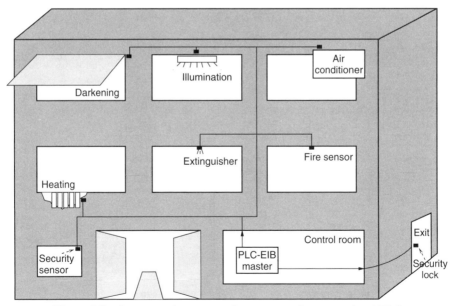

Figure 2.8 Structure of an automation system using narrowband PLC

any wall socket or integrated in any device connected to the electrical installation. This ensures communications between all parts of an internal electrical network. Nowadays, the PLC modems using FSK achieve data rates up to 1200 bps [Dost01].

As it is specified in CENELEC standard, power supply utilities can use band A for the realization of so-called *energy-related services*. In this way, a power utility can use PLC to realize internal communications between its control center and different devices, ensuring remote control functions, without building extra telecommunications network or buying network resources at a network provider (Fig. 2.9). Simultaneously, PLC can be used for remote reading of a customer's meter units, which additionally saves cost on the personnel needed for manual meter reading. Finally, PLC can also be used by the utilities for dynamic pricing (e.g. depending on the day time, total energy offer, etc.), as well as for observation and control of energy consumption and production. In the last case, especially, the utilities have been trying to integrate an increasing number of small power plants; for example, small hydroelectric power stations, wind plants, and so on. However, the small power plants are not completely reliable and their power production varies depending on the current weather conditions. Therefore, the regions that are supplied by the small plants should also be supplied from other sources if necessary. For this purpose, the utilities need a permanent communication between their system entities, which can be at least partly realized by PLC as well.

The building automation is a typical indoor application of the narrowband PLC systems, whereas the energy-related services are mainly (not only) indoor applications. In [BumiPi03], we find a very interesting example of an application of a PLC-based automation system in the outdoor area. In this case, a PLC-based airfield ground–lighting automation system is used

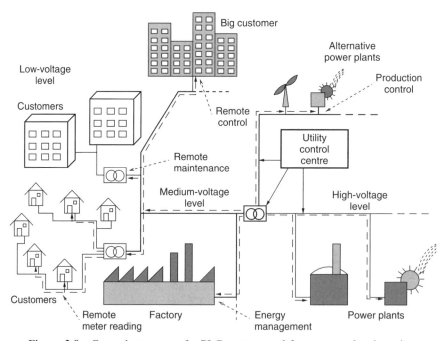

Figure 2.9 General structure of a PLC system used for energy-related services

for individual switching and monitoring of airfield lighting. The length of the airfields and accordingly the necessary communications networks in a large airport is very long (several kilometers). So, the narrowband PLC can be applied to save costs on building a separate communications network. This is also an example of PLC usage for the realization of so-called *critical automation services* with very high security requirements, such as the light control of ground aircraft movement in the airports.

2.2.5 Broadband PLC

Broadband PLC systems provide significantly higher data rates (more than 2 Mbps) than narrowband PLC systems. Where the narrowband networks can realize only a small number of voice channels and data transmission with very low bit rates, broadband PLC networks offer the realization of more sophisticated telecommunication services; multiple voice connections, high-speed data transmission, transfer of video signals, and narrowband services as well. Therefore, PLC broadband systems are also considered a capable telecommunications technology.

The realization of broadband communications services over powerline grids offers a great opportunity for cost-effective telecommunications networks without the laying of new cables. However, electrical supply networks are not designed for information transfer and there are some limiting factors in the application of broadband PLC technology. Therefore, the distances that can be covered, as well as the data rates that can be realized by PLC systems, are limited. A further very important aspect for application of broadband PLC is its Electromagnetic Compatibility (EMC). For the realization of broadband PLC, a significantly wider frequency spectrum is needed (up to 30 MHz) than is provided within CENELEC bands. On the other hand, a PLC network acts as an antenna becoming a noise source for other communication systems working in the same frequency range (e.g. various radio services). Because of this, broadband PLC systems have to operate with a limited signal power, which decreases their performance (data rates, distances).

Current broadband PLC systems provide data rates beyond 2 Mbps in the outdoor arena, which includes medium- and low-voltage supply networks (Fig. 2.7), and up to 12 Mbps in the in-home area. Some manufacturers have already developed product prototypes providing much higher data rates (about 40 Mbps). Medium-voltage PLC technology is usually used for the realization of point-to-point connections bridging distances up to several hundred meters. Typical application areas of such systems is the connection of local area networks (LAN) networks between buildings or within a campus and the connection of antennas and base stations of cellular communication systems to their backbone networks. Low-voltage PLC technology is used for the realization of the so-called "last mile" of telecommunication access networks. Because of the importance of telecommunication access, current development of broadband PLC technology is mostly directed toward applications in access networks including the in-home area. In contrast to narrowband PLC systems, there are no specified standards that apply to broadband PLC networks (Sec. 2.2.3).

2.3 PLC Access Networks

2.3.1 Structure of PLC Access Networks

The low-voltage supply networks consist of a transformer unit and a number of power supply cables linking the end users, which are connected to the network over meter units.

A powerline transmission system applied to a low-voltage network uses it as a medium for the realization of PLC access networks. In this way, the low-voltage networks can be used for the realization of the so-called "last mile" communications networks.

The low-voltage supply networks are connected to medium- and high-voltage networks via a transformer unit (Fig. 2.10). The PLC access networks are connected to the backbone communications networks (WAN) via a base/master station (BS) usually placed within the transformer unit. Many utilities supplying electrical power have their own telecommunications networks linking their transformer units and they can be used as a backbone network. If this is not the case, the transformer units can be connected to a conventional telecommunications network.

The connection to the backbone network can also be realized via a subscriber or a power street cabinet, especially if there is a convenient possibility for its installation (e.g. there is a suitable cable existing that can be used for this purpose at low cost). In any case, the communications signal from the backbone has to be converted into a form that makes possible its transmission over a low-voltage power supply network. The conversion takes place in a main/base station of the PLC system.

The PLC subscribers are connected to the network via a PLC modem placed in the electrical power meter unit (M, Fig. 2.10) or connected to any socket in the internal electrical network. In the first case, the subscribers within a house or a building are connected to the PLC modem using another communications technology (e.g. DSL, WLAN). In the second case, the internal electrical installation is used as a transmission medium that leads to the so-called *in-home PLC solution* (Sec. 2.3.2).

The modem converts the signal received from the PLC network into a standard form that can be processed by conventional communications systems. On the user side, standard communications interfaces (such as Ethernet and ISDN S_0) are usually offered. Within a house, the transmission can be realized via a separated communications network or via an internal electric installation (in-home PLC solution). In this way, a number of communications devices within a house can also be connected to a PLC access network.

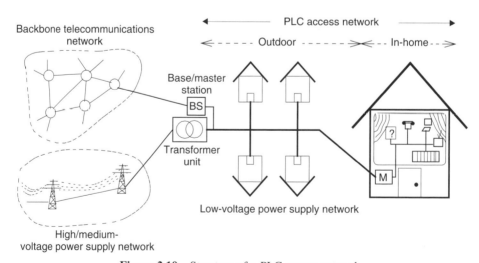

Figure 2.10 Structure of a PLC access network

2.3.2 In-home PLC Networks

In-home PLC (indoor) systems use internal electrical infrastructure as transmission medium. It makes possible the realization of PLC local networks within houses, which connect some typical devices existing in private homes; telephones, computers, printers, video devices, and so on. In the same way, small offices can be provided with PLC LAN systems. In both cases, the laying of new communications cables at high cost is avoided.

Nowadays, automation services are becoming more and more popular not only for their application in the industrial and business sectors and within large buildings, but also for their application in private households. Systems providing automation services like security observation, heating control, automatic light control have to connect a big number of end devices such as sensors, cameras, electromotors, lights, and so on. Therefore, in-home PLC technology seems to be a reasonable solution for the realization of such networks with a large number of end devices, especially within older houses and buildings that do not have an appropriate internal communication infrastructure (Sec. 2.2.4).

Basically, the structure of an in-home PLC network is not much different from the PLC access systems using low-voltage supply networks. There can also a base station that controls an in-home PLC network, and probably connects it to the outdoor area (Fig. 2.11). The base station can be placed with the meter unit, or in any other suitable place in the in-home PLC network. All devices of an in-home PLC network are connected via PLC modems, such as the subscribers of a PLC access network. The modems are connected directly to the wall power supply sockets (outlets), which are available in the whole house/flat. Thus, different communications devices can be connected to the in-home PLC network wherever wall sockets are available.

An in-home PLC network can exist as an independent network covering only a house or a building. However, it excludes usage and control of in-home PLC services from a distance. On the other hand, a remote controlled in-home PLC system is very comfortable for the realization of various automation functions (e.g. security, energy management, see

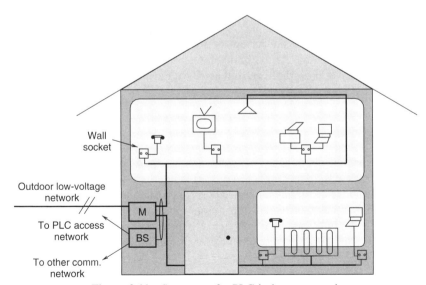

Figure 2.11 Structure of a PLC in-home network

Sec. 2.2.4). Also, connection of an in-home PLC network to a WAN communication system allows the usage of numerous telecommunications services from each electrical socket within a house.

In-home PLC networks can be connected not only to a PLC access system but also to an access network realized by any other communications technology. In the first case, if the access network is operated by a power utility, additional metering services can be realized; for example, remote reading of electrical meter instruments saves the cost of manual reading, or energy management, which can be combined with an attractive tariff structure. On the other hand, an in-home PLC network can be connected to the access networks provided by different network operators as well. Thus, the users of the in-home network can also profit from the liberalized telecommunications market.

On the other hand, there are also other cost-effective communications systems for the realization of the broadband in-home networks. Wireless LAN (WLAN) systems are already available on the market, providing transmission data rates beyond 20 Mbps (Sec. 2.1.3). So, in contrast to the in-home PLC, WLAN allows the mobile usage of telecommunications services, such as cordless telephony, and more convenient handles with various portable communication devices. Nowadays, WLAN components with significantly improved performance become cheaper making the penetration of the in-home PLC technology more difficult.

2.3.3 PLC Network Elements

As mentioned above, PLC networks use the electrical supply grids as a medium for the transmission of different kinds of information and the realization of various communications and automation services. However, the communications signal has to be converted into a form that allows the transmission via electrical networks. For this purpose, PLC networks include some specific network elements ensuring signal conversion and its transmission along the power grids.

2.3.3.1 Basic Network Elements

Basic PLC network elements are necessary for the realization of communication over electrical grids. The main task of the basic elements is signal preparation and conversion for its transmission over powerlines as well as signal reception. The following two devices exist in every PLC access network:

- PLC modem
- PLC base/master station.

A PLC modem connects standard communications equipment, used by the subscribers, to a powerline transmission medium. The user-side interface can provide various standard interfaces for different communications devices (e.g. Ethernet and Universal Serial Bus (USB) interfaces for realization of data transmission and S_0 and a/b interfaces for telephony). On the other side, the PLC modem is connected to the power grid using a specific coupling method that allows the feeding of communications signals to the powerline medium and its reception (Fig. 2.12).

The coupling has to ensure a safe galvanic separation and act as a high pass filter dividing the communications signal (above 9 kHz) from the electrical power (50 or 60 Hz).

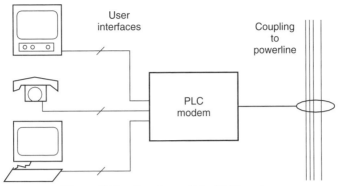

Figure 2.12 Functions of the PLC modem

To reduce electromagnetic emissions from the powerline, the coupling is realized between two phases in the access area and between a phase and the neutral conductor in the indoor area [Dost01]. The PLC modem implements all the functions of the physical layer including modulation and coding. The second communications layer (data link layer) is also implemented within the modem including its MAC (Medium Access Control) and LLC (Logical Link Control) sublayers (according to the OSI (Open Systems Interconnection) reference model, see for example [Walke99]).

A PLC base station (master station) connects a PLC access system to its backbone network (Fig. 2.10). It realizes the connection between the backbone communications network and the powerline transmission medium. However, the base station does not connect individual subscriber devices, but it may provide multiple network communications interfaces, such as xDSL, Synchronous Digital Mierarch (SDH) for connection with a high-speed network, WLL for wireless interconnection, and so on. (Fig. 2.13). In this way, a PLC base station can be used to realize connection with backbone networks using various communication technologies.

Usually, the base station controls the operation of a PLC access network. However, the realization of network control or its particular functions can be realized in a distributed manner. In a special case, each PLC modem can take over the control of the network operation and the realization of the connection with the backbone network.

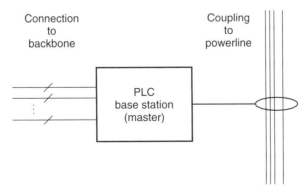

Figure 2.13 Function of the PLC base station

2.3.3.2 Repeater

In some cases, distances between PLC subscribers placed in a low-voltage supply network and between individual subscribers and the base station are too long to be bridged by a PLC access system. To make it possible to realize the longer network distances, it is necessary to apply a repeater technique. The repeaters divide a PLC access network into several network segments, the lengths of which can be overcome by the applied PLC system. Network segments are separated by using different frequency bands or by different time slots (Fig. 2.14). In the second case, a time slot is used for the transmission within the first network segment and another slot for the second segment.

In the case of frequency-based network segmentation, the repeater receives the transmission signal on the frequency f_1, amplifies and injects it into the network, but on the frequency f_2. In the opposite transmission direction, the conversion is carried out for frequency f_2 to f_1. Depending on applied transmission and modulation methods, the repeater function can include demodulation and modulation of the transmitted signal as well as its processing on a higher network layer. However, a repeater does not modify the contents of the transmitted information, which is always transparently transmitted between the network segments of an entire PLC access system (Fig. 2.15).

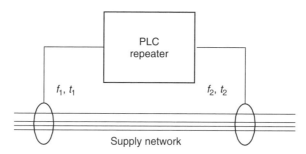

Figure 2.14 Function of the PLC repeater

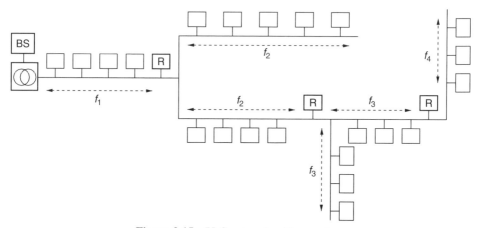

Figure 2.15 PLC network with repeaters

In a first network segment, between a base station placed in the transformer unit and the first repeater, the signal is transmitted within the frequency spectrum f_1. Another frequency range (f_2) has to be applied in the second network segment. Independent of the physical network topology, the signal is transmitted along both network branches. Theoretically, frequency range f_1 could be used again within the third network segment. However, if there is an interference between signals from the first segment, a third frequency range f_3 has to be applied to the third network segment and frequency f_4 to the fourth segment.

However, there is a limited frequency spectrum that can be used by the PLC technology (approximately up to 30 MHz), which is (or will be) specified by the regulatory bodies. So, with the increasing number of different frequency ranges, the common bandwidth is divided into smaller portions, which significantly reduces the network capacity. Therefore, a frequency plan for a PLC access network has to provide usage of as low a number of frequencies as possible. Application of the repeaters can extend network distances that are realized by the PLC technology. However, the application of repeaters also increases the network costs because of the increasing equipment and installation costs. Therefore, the number of repeaters within a PLC access network has to be kept as small as possible.

2.3.3.3 PLC Gateway

There are two approaches for the connection of the PLC subscribers via wall sockets to a PLC access network:

- Direct connection
- Indirect connection over a gateway.

In the first case, PLC modems are directly connected to the entire low-voltage network and with it to the PLC base station as well (Fig. 2.16). There is no division between the outdoor and indoor (in-home) areas, and the communications signal is transmitted through the power meter unit. However, the features of indoor and outdoor power supply networks are different, which causes additional problems regarding characteristics of PLC transmission channel and electromagnetic compatibility problems (as is explained later in

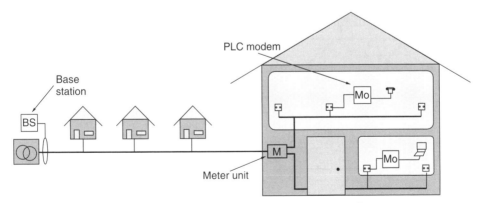

Figure 2.16 Direct connection of the PLC subscribers

the book). Therefore, the indirect connection using a gateway is a frequently used solution for the direct connection of the wall sockets to entire PLC access networks.

A gateway is used to divide a PLC access network and an in-home PLC network. It also converts the transmitted signal between the frequencies that are specified for use in the access and in-home areas. Such a gateway is usually placed near the house meter unit (Fig. 2.17). However, a PLC gateway can provide additional functions that ensure a division of the access and in-home areas on the logical network level too. Thus, PLC modems connected within an in-home network can communicate internally without information flow into the access area. In this case, a PLC gateway serves as a local base station that controls an in-home PLC network coordinating the communication between internal PLC modems and also between internal devices and a PLC access network (see Sec. 2.3.2).

Generally, a gateway can also be placed anywhere in a PLC access network to provide both signal regeneration (repeater function) and network division on the logical level. In this way, a PLC can be divided into several subnetworks that use the same physical transmission medium (the same low-voltage network), but exist separately as a kind of virtual network (Fig. 2.18). Both gateways (G) operate as PLC repeaters converting the transmission signal between frequencies f_1 and f_2 (or time slots t_1 and t_2), as well as between f_2 and f_3 (or t_2 and t_3). Additionally, the gateways control the subnetworks II and III, which means that internal communication within a subnetwork is taken over by a responsible gateway and does not affect the rest of a PLC access network, similar to that within in-home networks using a gateway. The communication between a member of a subnetwork and the base station is possible only over a responsible gateway. However, the network can be organized so that the base station directly controls a number of subscribers (subnetwork I).

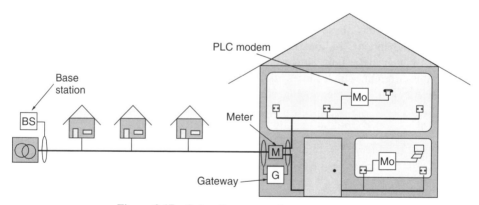

Figure 2.17 Subscriber connection over gateway

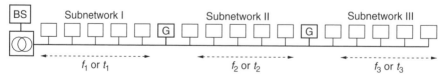

Figure 2.18 Gateways in the PLC access network

The gateways are connected to the network in the same way as the repeaters (Fig. 2.14). Also, an increasing number of gateways within a PLC access network reduces its network capacity and causes higher costs. However, where the repeaters provide only a simple signal forwarding between the network segments, the gateways can provide more intelligent division of the available network resources, ensuring better network efficiency as well.

2.3.4 Connection to the Core Network

A PLC access network covers the so-called "last mile" of the telecommunications access area. This means that the last few hundred meters of the access networks can be realized by PLC technology applied to the low-voltage supply networks. On the other hand, PLC access networks are connected to the backbone network through communications distribution networks, as is shown in Fig. 2.19. In general, a distribution network connects a PLC base station with a local exchange office operated by a network provider.

As mentioned in Sec. 2.1, the application of PLC technology should save the costs on building new telecommunications networks. However, the PLC access network has to be connected to the WAN via backbone networks that cause additional costs as well. Therefore, a PLC backbone network has to be realized with the lowest possible investments to ensure the competitiveness of PLC networks with other access technologies.

2.3.4.1 Communications Technologies for PLC Distribution Networks

The cheapest solution for the realization of the connection between a PLC access and the backbone network is usage of communications systems that are available in the application area. Some transformer units are already connected to a maintenance network via standard communications cables (copper lines). Originally, these connections were provided for the realization of remote control functions and internal communications between a control center of the supply network and the maintenance personnel and equipment. However, they can be used for the connection of PLC networks to the backbone by applying one of the DSL technologies (Sec. 2.1.3).

During the last decade, many electrical utilities realized optical communications networks along their supply lines, which can be applied for connection to the backbone as well. In this case, an access network consists of an optical and a PLC network part (Fig. 2.19), which leads to a hybrid solution similar to HFC networks (Hybrid Fiber Coax), in which an optical distribution network connects CATV access networks to WAN. A further solution for the realization of the backbone connection is application of PLC technology in medium-voltage supply networks (Sec. 2.3.5), which are, in any case, connected to the low-voltage networks.

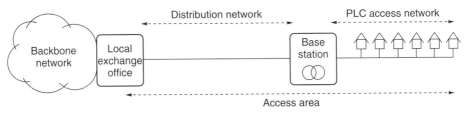

Figure 2.19 Connection to the backbone network

Application of a particular communications technology to the PLC backbone connection depends also on technical opportunities of a network provider operating PLC access networks. Usage of existing communication systems, of a supply utility or an independent network provider, is always a privileged solution. Generally, there are the following possibilities for the realization of the connection to the core network:

- Usage of the existing or new cable or optical networks
- Realization of wireless distribution networks; e.g. WLL (Sec. 2.1.2), application of satellite technology, and so on.
- Application of PLC technology in the MV supply networks.

Communications technology applied to the PLC distribution networks has to ensure transmission of all services that are offered in the PLC access networks. Also, PLC backbone networks must not be a bottleneck in the common communications structure between PLC subscribers and the backbone network. Therefore, an applied backbone technology has to provide enough transmission capacity (data rates) and realization of various Quality of Service (QoS) guarantees.

2.3.4.2 Topology of the Distribution Networks

A reasonable solution for the connection of multiple PLC access networks, placed within a smaller area, is the realization of a joint distribution network connecting a number of PLC networks, as shown in Fig. 2.20. The distribution networks can be realized in different topologies independent of applied communications technology (bus, star, ring). A chosen network topology has to ensure a cost-effective, but also a reliable, solution (including a redundancy in the case of failure), and this depends primarily on the location of PLC access networks in a considered area and on the position of the local exchange office (Fig. 2.19).

Bus network topology is one of the possible solutions that can be realized at low costs within adequate application areas (Fig. 2.20). However, the cost factor is not the single criterion for the decision about the topology of the distribution network. A very important criterion is the network reliability in the case of link failures. So, in the bus topology,

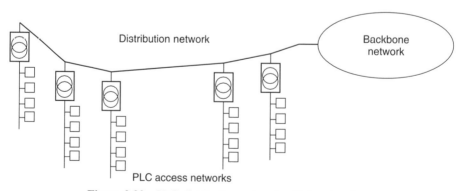

Figure 2.20 PLC distribution network with bus topology

if a link between two PLC access networks breaks down, all access networks placed behind the failed link are also disconnected from the WAN. Therefore, meshed network topologies have to be considered for application in the PLC distribution networks. A possible solution is a network with a star topology connecting each PLC access network separately (Fig. 2.21).

The star network topology is adequate for application of DSL technology in PLC distribution networks. However, failure of a single link in the star network disconnects only one PLC access network and there is no possibility for the realization of an alternative connection of the affected PLC access network to the backbone over a redundant transmission link. Therefore, the application of ring network topology (Fig. 2.22) seems to be a reasonable solution for increasing the network reliability. In the case of a failure in a single link between the ring nodes, there is always an opportunity for realization of the alternative transmission paths. Of course, reorganization of the transmission paths

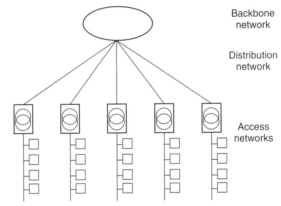

Figure 2.21 Star distribution network

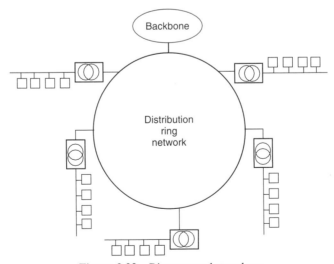

Figure 2.22 Ring network topology

between the PLC access networks and the backbone has to be done automatically within a relatively short time interval (maximum several seconds). Thus, applied transmission technology in the backbone networks has to support the implementation in a ring network structure (e.g. Distributed Queue Dual Bus (DQDB), Fiber Distributed Data Interface (FDDI)).

Finally, the topology of a PLC distribution network can also be a combination of any of the three basic network structures presented above. However, the choice for a network topology depends on several factors, among others:

- Used communications technology causing a specific network topology,
- Availability of a transmission medium within the application area,
- Possibility of the realization of reliable distribution networks
- Geographical structure and distribution of PLC access networks and a local exchange office.

2.3.4.3 Managing PLC Access Networks

An efficient control of the PLC access networks has to be done from one or a very small number of management centers providing an economically reasonable solution. However, PLC access networks belonging to a network or service provider can exist in a geographically wider area or a number of PLC networks can be distributed in several geographically separated regions. Therefore, it is important to optimize the management system that is used for the control of multiple PLC access networks (Fig. 2.23).

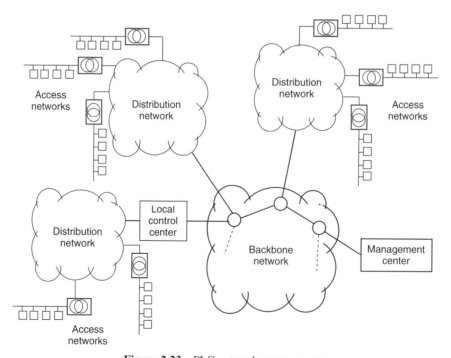

Figure 2.23 PLC network management

Management of a PLC access network includes configuration and reconfiguration of all its elements (base station, modems, repeaters and gateways) depending on the current network status. The management functions can be done locally by the base station or gateways or by a management center using remote control functions. Local management is done automatically without any action of the management personnel. On the other hand, remote management provides both automatic and manual execution of control functions. Transmission of management information from and to the access networks has to be ensured over PLC distribution networks to avoid buildup of particular management communications systems. An efficient management solution is the transfer of possibly more maintenance functions to the base stations and gateways placed in the access networks. However, management ability of PLC network elements increases the equipment costs. Therefore, the division of management functions between the network elements and a central office is an optimization task as well.

Anyway, the basic network operation has to be ensured by PLC network elements themselves, without any action of a management center. Once the equipment is installed in a low-voltage network, a PLC network that provides a number of self-control and self-configuration procedures should operate without the aid of the maintaining personnel. PLC access networks can be operated with economical efficiency only if the need for manual network control is reduced, especially activities that are carried out directly on the network locations.

2.3.5 Medium-voltage PLC

Similar to the PLC access systems using low-voltage power supply networks as a transmission medium, the medium-voltage supply networks can also be used for the realization of various PLC services. Generally, the organization of the so-called *medium-voltage PLC* (MV PLC) is not different from the PLC in the low-voltage networks. Thus, the medium-voltage PLC networks include the same network elements (Sec. 2.3.3): PLC modems connecting the end users with the medium-voltage transmission medium, base station connecting a medium-voltage PLC network to the backbone, repeaters and gateways.

A medium-voltage electrical network usually supplies several low-voltage networks, as is mentioned in Sec. 2.2.2 and presented in Fig. 2.7. Accordingly, an MV PLC network can be used as a distribution network connecting a number of PLC access networks to the backbone. In this case, several PLC access networks are connected to the MV PLC distribution network with a network topology similar to the ring distribution network presented in Fig. 2.22.

However, the transmission features of the medium-voltage supply networks, considered for their application in communications, seem to be similar to the low-voltage networks. Even the transmission conditions in the medium-voltage networks are better than in the low-voltage networks used for the realization of PLC access networks; the data rates to be realized over MV PLC are expected to be not significantly higher than in the PLC access networks. Accordingly, if a MV PLC network is used to connect a higher number of PLC access networks to the core network, the transmission part over the medium-voltage power grids would be a bottleneck. Therefore, it is not expected that the MV PLC networks will be used for the interconnection of multiple PLC access networks (e.g. to connect more than two access networks). However, in the developing phase it is expected that PLC access networks connect a fewer number of end users and in this case, the MV networks can be used as a solution for the distribution network.

On the other hand, the MV PLC offers an opportunity for the realization of communications networks without the need for the laying of new communications cables in a wider covering area. So, a medium-voltage supply network can be used for the connection of multiple LAN within a campus in a common data network, as shown in Fig. 2.24.

In the same way, the MV PLC can be applied for the realization of various point-to-point connections, which can be used for interconnection between LAN, similar to the campus network shown in Fig. 2.24. Nowadays, the MV PLC is mainly applied for the realization of such point-to-point connections. An application of MV PLC is the connection of antennas for various radio systems. In this way, an antenna used for a wireless mobile system (see Fig. 2.2) can be connected to its base station via a medium-voltage supply network.

2.4 Specific PLC Performance Problems

In previous sections, it has been shown that PLC technology presents a cost-effective alternative for the realization of the access networks. On the other hand, electrical supply networks are not designed for communications and therefore, they do not represent a favorable transmission medium. In this section, we outline some specific performance problems limiting the application of PLC technology and present several solutions to overcome these problems. A more detailed consideration of the performance limitations as well as various technical solutions for PLC networks are presented in the following chapters of the book.

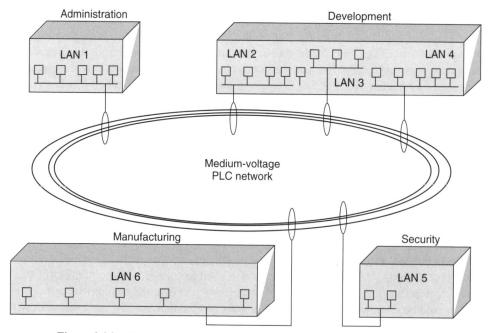

Figure 2.24 Structure of a campus communications network using MV PLC

2.4.1 Features of PLC Transmission Channel

The low-voltage supply networks are not designed for communications and accordingly, the transmission characteristics of powerline channels, are not favorable for data transfer. The powerline cables are divided in an asymmetric way (Fig. 2.25), having many irregular connections between network sections and customers and transitions between overhead and underground cables (Fig. 2.7). The cable transitions cause reflections and changing characteristic impedance [ZimmDo00a]. Additionally, a PLC network changes its structure (e.g. by adding new customers), especially in an in-home PLC network (Fig. 2.11) in which every switching event can change the network topology.

PLC networks are also characterized by multipath propagation because of numerous reflections caused by the joining of cables and their different impedances. This results in multipath signal propagation, with a frequency-selective fading. The most important effects influencing signal propagation are cable losses, losses due to reflections at branching points and mismatched endings of the cables as well as selective fading [Dost01, ZimmDo00a, Zimm00, ZimmDo02].

Attenuation in PLC networks depends on the line, length and changing characteristic impedance of the transmission line. Numerous measurements (e.g. [Dost01, ZimmDo00a]) have shown that the attenuation in powerlines is acceptable in relatively short cables (approximately up to 200–300 m), but is very bad in longer cables. Therefore, longer PLC networks are expected to be equipped with the repeater technique (Sec. 2.3.3).

A detailed description of the transmission features of the electrical supply networks is given in Sec. 3.2.

2.4.2 Electromagnetic Compatibility

The low-voltage supply networks used as a transmission medium for PLC access systems act as an antenna producing electromagnetic radiation. On the other hand, the PLC systems that allow realization of broadband access networks use a frequency spectrum of up to

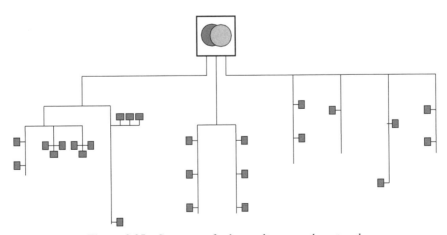

Figure 2.25 Structure of a low-voltage supply network

30 MHz, as mentioned in Sec. 2.2.5. This frequency range is reserved for various radio services and they may be disturbed by PLC systems. In the first place, the operation of various shortwave radio services, such as amateur radio, different public services, military and even very sensitive services like flight control, can be negatively affected by the disturbances coming from the PLC networks.

The regulatory bodies specify the limits for electromagnetic emission that is allowed to be produced by PLC systems operating out of the frequency range defined by CENELEC standard [palas00]. In Germany, NB30 directions [NB30] define very low radiation limits for systems operating in the frequency range up to 30 MHz. Accordingly, the PLC networks have to operate with a limited signal power to keep NB30 directions. The following two solutions are proposed for the specification of the frequency ranges to be used by the PLC:

- *Chimney Approach*: A total frequency spectrum of approximately 7.5 MHz in the frequency range between 1 and 30 MHz may be principally used for PLC. However, the spectrum does not continuously provide frequency ranges that are allowed to be used by PLC. In the allowed ranges, PLC still has to operate under specified radiation limits.
- *General Radiation Limitation*: In the entire frequency spectrum (below 30 MHz), the maximum radiation fields are limited for all wireline telecommunications services (including DSL, CATV and PLC).

In both cases, PLC systems have to ensure very low values regarding the electromagnetic emission and, accordingly, operate with the limited signal power. This problem is increased as the power cabling is neither schilded nor twisted pair. Various EMC aspects are considered in detail in Sec. 3.3.

2.4.3 Impact of Disturbances and Data Rate Limitation

Because of the limited signal power, PLC networks become more sensitive to the disturbances and are not able to span longer distances for ensuring a sufficient transmission capacity. The disturbances from the PLC network environment are caused by other services (such as shortwave radio) operating in the frequency range below 30 MHz (Fig. 2.26). There are also disturbances coming from the PLC network itself; heavy machines, such as electromotors, which could be connected to the low-voltage network or can exist near the PLC network, TV and computer monitors as well as disturbance impulses caused by on/off switching of appliances and phase angle control devices. Finally, disturbances can be caused by neighboring PLC networks as well. A detailed noise classification and a description of disturbance models can be found in Sec. 3.4.

Well-known error-handling mechanisms can be applied to the PLC systems to solve the problem of transmission errors caused by the disturbances (e.g. FEC and ARQ, see Sec. 4.3). Forward error correction (FEC) mechanisms can recover the original contents of a data unit in spite of the disturbance influence. However, the application of FEC mechanisms consumes an additional part of the transmission capacity because of the overhead needed for the error correction. Usage of ARQ (Automatic Repeat reQuest) mechanisms provides retransmission of defective data units consuming a part of the transmission capacity and introducing extra transmission delays too.

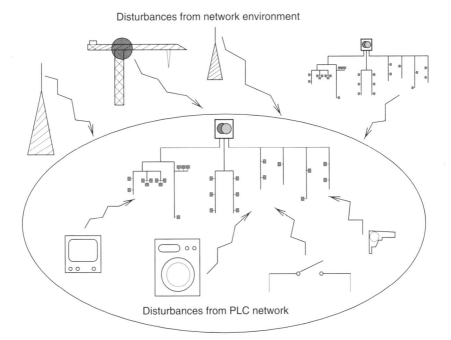

Figure 2.26 The influence of various disturbance sources

Application of error-handling mechanisms is needed in PLC networks because of the inconvenient disturbance behavior. On the other hand, data rates provided by PLC systems are limited because of the electromagnetic compatibility (EMC) requirements. So, currently offered PLC systems have maximum net data rates of 2 to 4 Mbps. Therefore, PLC networks have to operate with low data rates additionally decreased by the application of error-handling mechanisms. On the other hand, PLC access networks connect a number of subscribers who use a low-voltage supply network as a transmission medium (Fig. 2.27), which additionally decreases the available data rate.

As mentioned in Sec. 2.3, a PLC access network uses a low-voltage supply network to connect a number of PLC subscribers to the base station, which ensures connection to the wide area network. Thus, a PLC network represents a shared transmission medium used by all subscribers independently. Accordingly, the capacity of PLC networks is furthermore reduced.

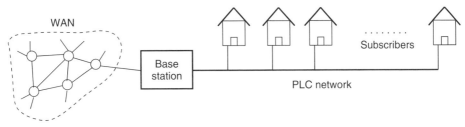

Figure 2.27 Shared transmission medium in PLC access networks

2.4.4 Realization of Broadband PLC Transmission Systems

Within the PLC systems, data transfer is carried out in a channel characterized by frequency-selective phenomena, the presence of echoes, impulsive and colored noise with the superposition of narrowband interferences. This requires that the modulation scheme adopted for PLC must effectively face such a hostile environment. DSS (Direct Sequence Spread Spectrum) and OFDM techniques are considered as candidates for future broadband PLC networks [Dost01, Zimm00, TachNa02, DelFa01].

The spread spectrum has the advantage of robustness to narrowband interferences, the possibility for the realization of CDMA (Code Division Multiple Access), and operation with a low power spectrum density reducing the EMC problems [Dost01, SchuSc00]. However, DSS has a low spectral efficiency and a low-pass characteristic, and it is sensitive to frequency-selective fading. Therefore, there is a need for complex signal equalization in point-to-multipoint connections, such as in PLC access networks, depending on the length of the network sections for individual connections [Dost01, Zimm00].

On the other hand, the OFDM technique allows a great reduction in the channel equalizer complexity and an increase in the resistance to the signal distortions. A feature of OFDM to use a frequency spectrum in a selective way is suitable for the avoidance of frequency ranges disturbed by narrowband interferences and to get around critical frequencies specified by the regulators (Sec. 2.4.2). Orthogonality, provided by OFDM allows spectral overlapping leading to outstanding efficiency, which is about twice as good as single-carrier broadband systems [Dost01]. Moreover, bit loading techniques, applied to OFDM subcarriers, make it possible to achieve a capacity very close to the theoretical limits of a transmission medium [DelFa01, SchuSc00]. For that reason, OFDM is considered as the favorite candidate for application in broadband PLC networks.

A detailed description of the system requirements for the realization of broadband PLC networks and proposed transmission and modulation schemes for PLC is presented in Sec. 4.2.

2.4.5 Performance Improvement by Efficient MAC Layer

Because of the competition in the telecommunications markets, network providers using PLC technology have to be able to offer attractive telecommunications services. In other words, PLC systems have to compete with other access technologies (Sec. 2.1) and to offer a satisfactory QoS. At the same time, PLC access networks have to be economically efficient as well. For these reasons, PLC access systems have to provide a very good network utilization of the shared transmission medium (Fig. 2.27) and, simultaneously, a satisfactory QoS. Both requirements can be achieved by the application of efficient MAC layer. The task of MAC layer is to organize the medium access between multiple subscribers using various telecommunications services.

Currently, there are no specifications or standards considering the MAC layer and protocols for PLC network. In spite of a rapid development of PLC technology during the last few years, there is also a limited number of published research works in this area. The manufacturers of PLC equipment apply their own protocol solutions, which differ between various PLC products.

A MAC layer specifies a multiple access scheme, a resource sharing strategy (MAC protocol) and mechanisms for traffic control in a network. The most widely applied access scheme in recent broadband PLC networks is TDMA (Time Division Multiple Access). Because of the disturbances, data packets can be segmented into smaller data units whose size is chosen according to the length of a time slot specified by the TDMA scheme. Thus, if a disturbance occurs, only erroneous data segments are retransmitted, consuming a smaller network capacity. The data segmentation ensures fine network granularity and an easier provision of QoS guarantees.

An effective solution for avoiding the influence of narrowband disturbances is to apply the FDMA method, in which particular frequencies can be switched off if they are affected by the disturbances. Therefore, a TDMA/FDMA combination seems to be a reasonable solution for PLC networks as well. Application of the FDMA in an OFDM-based PLC system leads to the so-called OFDMA (OFDM Access) scheme, which can also be combined with the TDMA building of an OFDMA/TDMA system.

MAC protocols for PLC systems have to achieve a maximum utilization of the limited network capacity and realize time-critical telecommunication services. This can be ensured by reservation of bandwidth, which allows particular QoS guarantees needed for various services. This is ensured by the so-called *reservation MAC protocols*. Besides the reservation MAC protocols, variations of the CSMA/CA (Carrier Sense Multiple Access with Collision Avoidance) are also widely applied in PLC access networks; for example, IEEE 802.11 MAC protocol specified for WLAN is used in PLC and can provide realization of various QoS guarantees.

Because of the asymmetric and changing nature of data traffic in the access area, dynamic duplex schemes are used in PLC access networks. This allows the optimal utilization of the network resources, in both downlink and uplink transmission directions according to the current load situation. However, the relatively small PLC network capacity makes it difficult for the simultaneous provision of a required QoS for a high number of subscribers. Therefore, PLC systems have to implement traffic scheduling strategies, including connection admission control (CAC), to limit the number of active subscribers ensuring a satisfactory QoS for currently admitted connections. In the same way, a part of the network resources has to be reserved for capacity reallocation in case of disturbances.

A description of the MAC layer and its protocols for their application in broadband PLC access networks can be found in Chapter 5. A comprehensive performance analysis of reservation MAC protocols for PLC is presented in Chapter 6.

2.5 Summary

Present powerline communications systems, using electrical power grids as transmission medium, provide relatively high data rates (beyond 2 Mbps). PLC can be applied to high-, medium- and low-voltage supply networks as well as within buildings. However, PLC technology is nowadays mainly used for access networks and in-home communications networks. This is because of the high cost of the access networks (about 50% of the investments in network infrastructure are needed for the access area) and the liberalization of the telecommunications market in many countries.

Broadband PLC systems applied to the telecommunications access area represent an alternative communications technology for the realization of the so-called "last mile" networks. PLC access networks cover the last few hundred meters of a communications network directly connecting the end customers. PLC subscribers are connected to the network via PLC modems that ensure data transfer over low-voltage supply grids. On the other hand, a PLC network is connected to the backbone network via a base station. Thus, build up of new access networks can be avoided by the usage of broadband PLC technology.

Power supply networks are not designed for communications and they do not present a favorable transmission medium. The PLC transmission channel is characterized by a large, frequency-dependent attenuation, changing impedance and fading as well as a strong influence of noise. On the other hand, broadband PLC networks have to operate in a frequency spectrum up to 30 MHz, which is used by various radio services too. Therefore, the regulatory bodies specify very strong limits regarding the electromagnetic emission from PLC networks to the environment. As a consequence, PLC networks have to operate with a limited signal power, which reduces network distances and data rates, and also increases sensitivity to disturbances.

To reduce the negative impact of powerline transmission medium, PLC systems apply efficient modulation techniques, such as spread spectrum and OFDM. The problem of disturbances can be solved by well-known error-handling mechanisms (e.g. FEC, ARQ). However, their application consumes a certain portion of the PLC network capacity due to overhead and retransmission. The PLC bandwidth is shared by the subscribers and therefore, any reduction of capacity due to protocol overhead should be minimized. At the same time, PLC systems have to compete with other access technologies and offer a big palette of telecommunication services with a satisfactory QoS. Both, good network utilization and provision of QoS guarantees can be achieved by an efficient MAC layer.

3

PLC Network Characteristics

In this chapter, we describe the characteristics of PLC networks using low-voltage power supply networks as a transmission medium. The low-voltage networks are characterized by their particular topology as well as by specific features if the supply networks are used as a transmission medium for communications. On the other hand, a PLC access network acts as an antenna producing electromagnetic emission, which disturbs other communications services operating in the same frequency range (up to 30 MHz). As a consequence, PLC systems have to operate with a limited signal power that makes them sensitive to disturbances. PLC networks are affected by disturbances from the network environment and also from the low-voltage network itself. In this chapter, the following four PLC specific characteristics are considered: network topology in different realizations of PLC access networks, specific features of the PLC transmission channel (low-voltage supply network), EMC (electromagnetic compatibility) issue, and the characteristics of noise that cause the disturbances in PLC networks.

3.1 Network Topology

The topology of a PLC access network is given by the topology of the low-voltage supply network used as a transmission medium. However, a PLC access network can be organized in different ways (e.g. different position of the base station, network segmentation, etc.), which can influence the network operation. In this section, we discuss various realizations of PLC access networks and their influence on the network topology and communication organization in the network. The impact of application of so-called additional network elements (repeaters and gateways, Sec. 2.3.3) on the network structure is also analyzed.

3.1.1 Topology of the Low-voltage Supply Networks

The low-voltage supply networks are realized by the usage of various technologies (different types of cables, transformer units, etc.) and are installed in accordance with the existing standards, which differ from country to country. We also find various kinds of cabling in the low-voltage networks. So, there are networks realized with overhead or underground powerlines, which have different transmission features (see Sec. 3.2), as well as combined overhead/ground cabling solutions. The topology of a low-voltage power supply network also differs from place to place and depends on several factors, such as:

- *Network Location* – A PLC network can be placed in a residential, industrial or business area. Furthermore, there is a difference between rural and urban residential areas. Industrial and business areas are characterized by a higher number of customers who are potential users of the PLC services. It is also expected that subscribers from business areas have different requirements than industrial subscribers and especially different than subscribers from residential areas. Similar differences can be recognized between urban and rural application areas as well.
- *Subscriber Density* – The number of users/subscribers in a low-voltage network as well as user concentration, vary from network to network. The subscribers can be mostly placed in single houses (low subscriber density), which is typical for the rural application areas, within small blocks including several individual customers (e.g. urban residential area), in buildings with a larger number of flats or offices, or within apartment or business towers (very high subscriber density), such as in big commercial quarters.
- *Network Length* – The longest distance between the transformer unit and a customer within a low-voltage network also differs from place to place. Usually, there is a significant network length difference between the urban and rural application areas.
- *Network Design* – Low-voltage networks usually consist of several network sections (branches) of varying number, which differs from network to network, as well.

Figure 3.1 shows a possible structure of a PLC network. There are generally several branches (network sections) connecting the transformer station with the end users. Each branch can have a different topology connecting a variable number of users. The users can be more or less concentrated, and they can be distributed in a symmetric or in an asymmetric way along the low-voltage network or along its branches. There is also a difference between the lengths of the branches. Both low-voltage networks and their branches have a physical tree topology.

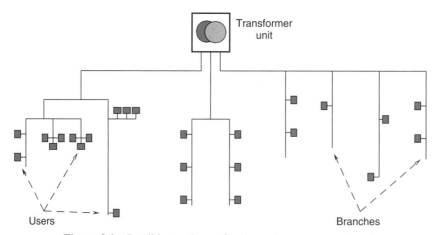

Figure 3.1 Possible topology of a low-voltage supply network

Low-voltage supply networks differ from each other and it is not possible to specify a typical network structure for them. However, it is possible to define some characteristic values, and to describe an average structure of a typical PLC network in accordance with the information from [Hooi98, HrasLe00, HrasHa01b] as follows:

- Number of users in the network: ~250 to 400
- Number of network sections: ~5
- Number of users in a network section: ~50 to 80
- Network length: ~500 m.

Note that the users of a supply network are only potential PLC subscribers and they do not have to use PLC services.

3.1.2 Organization of PLC Access Networks

The low-voltage supply networks, the topology of which is presented above, are used as a transmission medium for PLC access networks. However, there are several possibilities for organization of the PLC access systems using the same supply network or using the multiple low-voltage networks. In the following sections, we consider several possibilities for positioning a PLC base station in the network, network segmentation consisting of multiple PLC subnetworks, usage of multiple supply networks for realization of a PLC access network, and PLC networks with applied repeater and gateway techniques.

3.1.2.1 Position of the Base Station

As mentioned in Sec. 2.3, there is a main/base station in a PLC access network. The base station connects the PLC access system to the backbone network (wide area network (WAN)) and accordingly, it has a central place in the PLC network structure. There are the following two possibilities for placement of the base station:

- The base station is placed in the transformer unit with the connection to the WAN, and the PLC access network keeps the topology of the low-voltage supply network (Fig. 3.2).
- The base station is situated on the premises of a PLC subscriber or any other place in the network (e.g. power street cabinet). The topology of the PLC network changes and it can vary from the topology of the supply network, as shown in Fig. 3.3.

As mentioned above, the base station does not have to be placed within the transformer unit. Its position in a PLC access network depends primarily on the possibility of connecting the base station with the backbone network. Accordingly, the base station can be placed on the premises of a PLC subscriber (e.g. if there exists a convenient possibility for the WAN connection) or within street cabinets, which, in a general case, exist in different places within a low-voltage network. The street cabinets are usually equipped with a communication cable, originally provided for realization of the remote maintenance and internal utility communications, which can be used for the connection to the backbone as well.

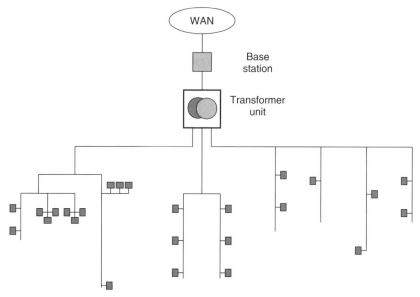

Figure 3.2 PLC network with the base station in the transformer unit

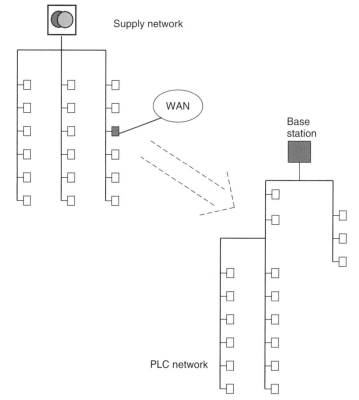

Figure 3.3 Topology of a PLC access network and corresponding low-voltage network

If the base station is not placed in the transformer unit, the central point (connection point to the backbone) of the PLC network moves to another place in the network. However, the position of the base station can move only along existing power supply grids (Fig. 3.3). This can only cause varying distances between the base station and subscribers in various network realizations. Thus, the topology of the PLC access network always remains the same, keeping the same physical tree structure.

3.1.2.2 Network Segmentation

A PLC access network can be realized to include a whole low-voltage power supply network or to include only a part of a supply network. To reduce the number of users per PLC system and the network length, it is possible to divide the low-voltage network into several parts (e.g. one PLC system per network section). In this case, several PLC systems can work simultaneously in a low-voltage network. Fig. 3.4 presents a possible segmentation of the low-voltage supply network that consists of three network sections. Each network section has a base station that connects a number of subscribers of a separated PLC access network. So, there are three separate PLC access systems within the low-voltage network. In this way, the number of subscribers who share the available network capacity is reduced.

One result of the network segmentation in multiple PLC access systems is a reduced length of originated PLC networks operating in individual network sections. Accordingly, the transmission can be realized with a lower signal power, which is important because of the electromagnetic compatibility problem (EMC, Sec. 2.4.2, Sec. 3.3). There are also a smaller number of potential subscribers in a network section than in the whole supply network and the transmission capacity is shared by a smaller number of PLC subscribers. The network segmentation is not limited only to network sections/branches. Each part of a supply network could also be realized as a separate PLC access system. It causes a further decrease in network length and in the number of subscribers connected to a PLC access network. It can be concluded that individual PLC systems within a low-voltage network also keep the physical tree topology.

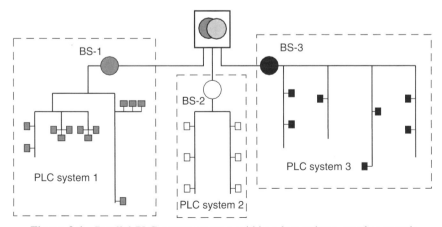

Figure 3.4 Parallel PLC access systems within a low-voltage supply network

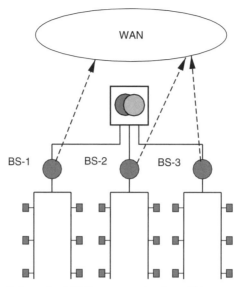

Figure 3.5 Independent PLC access networks within a supply network

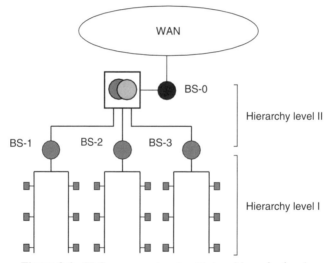

Figure 3.6 PLC access network with two hierarchy levels

Each of the individual PLC systems can be connected to the WAN separately (Fig. 3.4) representing independent PLC access networks (Fig. 3.5).

Another possibility for the connection to the core network is that the base stations use the supply network as a transmission medium for the connection to a central base station (BS-0, Fig. 3.6), which is connected to the backbone, thereby building a second network hierarchy. PLC networks with multiple hierarchy levels can be realized in the same manner, too. The base stations can share the PLC medium for communicating to the upper network level, or a separated frequency spectrum can be reserved for each base station for this communication. In both cases, there is a reduction of the available network capacity. Therefore, the realization of such hierarchical PLC access networks is not advantageous, and is therefore not expected.

However, if the distance is short between the base stations and the central point of an upper network hierarchy level, higher data rates can be realized in the upper network level (e.g. second level). If the data rate is sufficient to take on traffic load from all base stations simultaneously, there is no bottleneck in the upper network level and therefore, the realization of hierarchical PLC networks could make sense.

3.1.2.3 PLC over Multiple Low-voltage Networks

Low-voltage supply networks are very often interconnected, ensuring a redundancy in the energy supply system (Fig. 3.7). So, if a transformer unit malfunctions or is disconnected from the middle-voltage level, the supply can be realized over neighboring distribution networks and their transformer units. In normal cases, there is no current flow between two neighboring low-voltage networks. On the other hand, the designated interconnection points can be easily equipped to ensure transmissions of high-frequency signals used for communications. Accordingly, a PLC network can be realized to include multiple low-voltage networks. In this case, a base station connects PLC subscribers of all interconnected low-voltage networks to the WAN. Such networks covering multiple low-voltage supply systems keep the physical tree topology, as well.

In this way, a PLC access network can serve a larger area with subscribers from different low-voltage networks. However, the network capacity remains limited, allowing connection of a certain number of PLC subscribers to keep a required QoS in the network. On the other hand, the realization of PLC over multiple low-voltage networks is favorable for the first building phase of a PLC-based access network. Thus, in the first phase, whereas the number of PLC subscribers is expected to be small, a coverage area can be realized with less expenditure. Of course, with an increasing number of subscribers, the

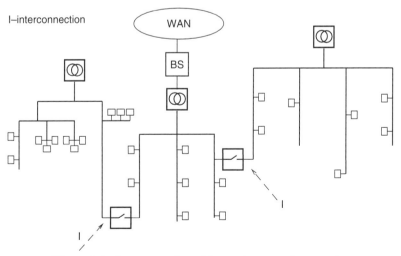

Figure 3.7 Interconnection of low-voltage supply networks

PLC network can be further developed to include a PLC system per low-voltage network or to include multiple PLC access systems within a low-voltage network.

3.1.2.4 Networks with Repeater and Gateway Technique

As mentioned in Chapter 2, the distance that can be spanned by PLC access networks ensuring reasonable data rates depends on the power of the injected signal. On the other hand, a higher signal power causes significant electromagnetic radiation into the PLC network environment. Therefore, PLC networks that overcome longer distances can offer very low data rates. However, realization of PLC access networks spanning longer distances and ensuring sufficient data rates is possible by application of a repeater technique.

Figure 3.8 presents an example of a PLC access network with repeaters. Distant parts of communications networks are connected to the base station via repeater devices that receive the signal and transmit the refreshed signals to another network segment. The repeaters operate bidirectionally and use either different frequencies or different time slots in the nearby network segments, as explained in Sec. 2.3.3. If it is necessary, the subscribers can be connected to the base station over multiple repeaters. Owing to the fact that a repeater only forwards the information flow between two nearby network segments, it can be concluded that a PLC access network using the repeater technique also keeps the physical tree network topology.

In the same way, a PLC access network can be divided into subnetworks by application of so-called PLC gateways (Sec. 2.3.3). In this case, each gateway controls a PLC network and realizes connection with a central base station. Thus, different from the repeaters, the gateways do not simply forward the data between the network segments and they additionally control the subnetworks. However, individual subnetworks also have the physical tree topology, such as in network realizations with multiple PLC access systems within a low-voltage supply network, described above.

Generally, an optional number of repeaters and gateways can be applied to a PLC access network dividing it into short network segments. However, a limiting factor for

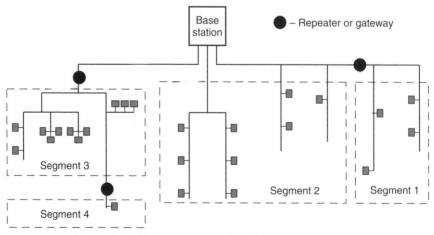

Figure 3.8 PLC access network with repeaters (gateways)

the realization of numerous short network segments within a PLC access network is the interference between the nearby segments. Therefore, a wider frequency spectrum has to be used and divided between network segments, which leads to the reduction of the common network capacity – such is the case in low-voltage networks with multiple PLC access systems.

The installation of the repeaters and gateways causes additional costs that can be avoided if the network stations, conveniently positioned in the network, also take the repeater or gateway functional. In the extreme case, each network station can operate simultaneously as a repeater, dividing a PLC network into very short network segments, which significantly decreases the necessary signal power and electromagnetic radiation (Solution proposed by the former company ONELINE, Barleben, Germany). However, network stations with the repeater function are more complex and their application requires a complicated management system to enable frequency or time-slot allocations within a PLC network. Furthermore, repeater devices cause additional propagation delays because of the processing time needed for the signal conversion. Therefore, the common number of repeaters, as well as gateways applied to a PLC access network is expected to be limited.

3.1.3 Structure of In-home PLC Networks

As was mentioned in Sec. 2.3, there are three possibilities for realization of the PLC in-home networks:

- An in-home electroinstallation is used as a simple extension of the PLC transmission medium provided by a low-voltage supply network.
- An in-home PLC network is connected via a gateway to an access network, which can be realized not only by a PLC system but also by any other access technology (e.g. DSL).
- An in-home PLC network exists as an independent system.

In the first case, the in-home electrical network is a part of a homogeneous PLC access network. A communications signal transmitted over a low-voltage network does not end up in the meter unit and it can also be transmitted through the in-home installation (Fig. 3.9). In this way, the connection to the PLC access system is available in each socket within the house. An internal electroinstallation, as an in-home part of the PLC access network, also keeps the same physical tree topology, as is recognized within low-voltage supply networks, too.

In-home PLC networks can also be connected over a gateway to any access network (Sec. 2.3). In this case, the gateway acts as a user on the site of the access network and as a main/base station for the in-home PLC network. If both access and in-home networks use PLC technology, the gateway is placed within the meter unit. This is also a point where all three current phases can be easily connected to each other, making PLC access available in each part of the internal electroinstallation. Accordingly, this is also a favorable place for the gateway if the access network is realized by other technology.

Independent in-home PLC networks include a base station that incorporates a master function for the entire home PLC system. It can be assumed that the base station of an independent in-home PLC network is also situated in the meter unit (Fig. 3.9). Independent

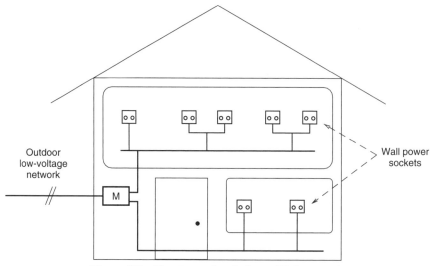

Figure 3.9 Topology of an in-home PLC network

of the kind of in-home PLC network, it keeps the physical tree topology, such as PLC access networks. Also, if the base station is moved to another place within the in-home PLC network (e.g. to a wall socket), the physical tree structure remains. However, the in-home networks are significantly shorter than the access networks, even if larger buildings are considered.

Some in-home PLC networks are organized in a decentralized manner, which leads to a network structure without PLC base station. This is usually the case in the independent in-home PLC networks, where the communication is organized by a negotiation between all network stations. However, the physical tree network structure can be recognized in those PLC networks, too.

3.1.4 Complex PLC Access Networks

In previous subsections, we have described network topologies of several PLC access networks realized in various ways. We considered the position of the PLC base station within a low-voltage supply network, network segmentation and interconnection, and PLC networks with repeater and gateway technique, as well as the in-home PLC networks. However, in a real environment, a PLC access network can be realized to include several of these features, building so-called complex PLC network structures.

In Fig. 3.10, we present a possible PLC network configuration covering multiple low-voltage networks and including different network elements. There are three supply networks in the example, each of them with a transformer unit supplying several branches, which connect variable numbers of users (potential PLC subscribers), and having also different user densities. The supply networks are interconnected (I) for the case in which a transformer unit falls out ensuring permanent supply to all users. In the normal case, the interconnection points are switched off, so there is no current flow between the supply networks. On the other hand, the interconnection points can be equipped to allow the transmission of high-frequency communications signals.

Because of the asymmetric division of the network users, there is a significantly higher number of PLC subscribers in the second supply network (Fig. 3.10). Therefore, the supply

PLC Network Characteristics 49

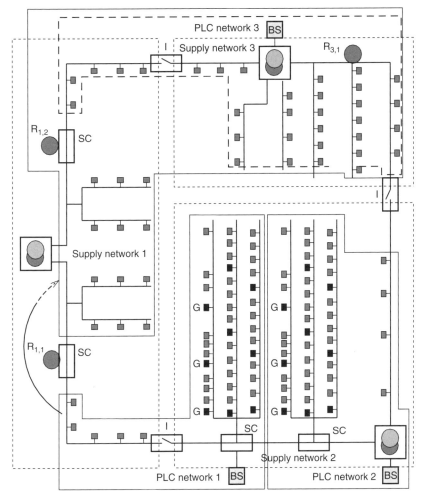

Figure 3.10 Example of a complex PLC access network

network is segmented into two PLC access systems, dividing PLC subscribers into two groups, and controlled by two separate base stations (BS). A base station is placed in the transformer unit and the second base station in a street cabinet (SC). Within the second supply network, the subscriber density is very high. Therefore, a number of gateways are installed to connect several subscriber groups to the base stations (e.g. a gateway for each apartment building with several PLC subscribers). The third PLC network covers supply network 3 and its base station is placed in the transformer unit. Within this network there is a need for repeater application to ensure communications with its distant subscribers ($R_{3,1}$).

It is assumed that the number of PLC subscribers in the first supply network is low or significantly lower than in the second and the third supply networks. Therefore, these subscribers can be connected to neighboring PLC access networks (networks 1 and 3) to save the costs for installation of an additional base station and its connection to the backbone network. Thus, PLC subscribers situated in supply network 1 are partly connected to the first and third PLC access networks and their base stations. Repeater $R_{1,2}$ ensures coverage of the subscribers, which are rather far from the base station of PLC network 3. In the usual case, repeater $R_{1,1}$ is not active (it is placed between areas of supply network 1 covered by PLC systems 1 and 3).

Traffic situation in access networks, such as PLC, varies during the day. The business subscribers are more active in the morning hours, whereas the private subscribers are more active in the evening. If we assume that the subscribers in supply network 3 are mainly private households (Fig. 3.10), and that there are several business customers in supply network 2, PLC access networks 1 and 2 are loaded higher during the day and PLC network 3 is loaded higher in the evening. Therefore, it would be reasonable to optimize the network load between PLC access systems, providing also better QoS in the network. So, to relieve PLC network 3, a part of PLC subscribers in the first supply network can be handed over to PLC access network 1. In this case, repeater $R_{1,1}$ becomes active, ensuring communications between the first base station and its coverage area in the first supply network, and repeater $R_{1,2}$ is switched off.

The change of PLC network configuration in an area with several PLC access systems can be carried out with a different dynamic, which depends on two factors: traffic load (as explained above) and transmission conditions in the network. However, to be able to react to the changing network conditions, the reconfiguration has to be carried out automatically. Thus, variation of the noise behavior in the network environment can lead to unfavorable transmission conditions that make communications with distant PLC subscribers difficult. In this case, the organization of repeaters and network interconnection can be changed to solve this problem. Even additional repeaters can be temporarily inserted in the network to overcome the problem. Note, that the subscriber network stations can also be designed to be able to take over the repeater function, which ensures the prompt insertion of additional repeaters.

3.1.5 Logical Network Models

As is considered for various PLC network realizations in Sec. 3.1.2, a PLC access network is connected to the backbone network over a base station. This connection exists in all realizations of PLC access systems independent of the position of the base station and the number of PLC subsystems within a low-voltage supply network. The communication between the subscribers and the WAN is carried out over the base station and it can be assumed that the internal communications between subscribers of a PLC network is also carried out via the base station as well.

For example, the data communication between subscribers within a PLC access network is carried out via an Internet server usually placed out of a PLC network. On the other hand, if the telephony service is considered, the connections are realized via a switching system also situated somewhere in the WAN. In accordance with this consideration, there are two transmission directions that can be recognized in a PLC network (Fig. 3.11):

- Downlink/downstream from the base station to the subscribers, and
- Uplink/upstream from the subscribers to the base station.

Information sent by the base station in the downlink direction is transmitted to all network subsections and is received by all subscribers in the network. In the uplink direction,

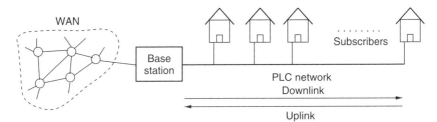

Figure 3.11 Logical PLC bus network structure

information sent by a PLC subscriber is received not only by the base station but also by all subscribers.

From the view of a higher network layer (e.g. MAC layer), a PLC access system can be considered as a logical bus network connecting a number of network stations with a base station, which provides communications with the WAN. Accordingly, the base station takes a central place in the communications structure of the bus network. The logical bus network does not include information about distances between the base station and the subscribers and between the subscribers themselves. This information is needed for the consideration of signal propagation delays in the network. For this purpose, a matrix can be defined to specify the distances between all stations in the network.

As analyzed in Sec. 3.1.2, the placement of the base station in PLC access networks does not change the network's physical tree structure. Accordingly, the logical bus network structure can be applied for consideration of higher network layers, as well. The same conclusion can be made if a low-voltage supply network is segmented into several PLC systems, or if multiple low-voltage networks are interconnected to build up a PLC access network. PLC in-home networks keep the same physical tree topology (Sec. 3.1.3) and accordingly, the logical bus network structure can be applied in this case, too.

As previously described, PLC access networks can be realized with repeaters. In this case, there is a number of network segments within a PLC system divided by the repeaters. Different frequency ranges or different time slots are used in different network segments, allowing their coexistence within a PLC access system. The repeaters convert the frequencies or the time slots between network segments without any impact on the data contents. Transmitted data units are simply passed between the network segments that ensure their continuous flow through the entire network. Therefore, the same logical bus network structure (Fig. 3.11) can also be used for the consideration of the higher network layers in PLC systems with the repeaters, as well as in networks with PLC gateways. If the network is divided in the time domain, the transmission delays caused by the time-slot transfer between the network segments have to be particularly taken into consideration.

In Sec. 3.1.4, we considered an example of a complex PLC access network containing several PLC access systems and base stations, repeaters and gateways, as well as covering multiple low-voltage supply networks. It was also concluded that the structure of multiple PLC access networks can change in the course of time because of changing conditions in the network. However, in spite of the interconnected low-voltage networks, every PLC access network has the physical tree structure (Fig. 3.10). Accordingly, the logical bus network can be applied for investigation of the higher network layers on each of the PLC access networks belonging to the complex structure. The change of the network structure also results in a similar physical topology with several tree networks. Thus, the logical bus model can be applied to each of the originated PLC access networks.

3.2 Features of PLC Transmission Channel

A transmission system in a telecommunications network has to convert the information data stream in a suitable form before this is injected in the communications channel (or medium). Like all other communications channels, the PLC medium introduces attenuation and phase shift on the signals. Furthermore, the PLC medium was at the beginning designed only for energy distribution, and for this reason several types of machines and appliances are connected to it. These activities on the power supply make this medium not adequate for information communications signals. Therefore, in this section we present an investigation of the PLC channel and its characteristics. Also, a PLC channel model is discussed, which describes the effect introduced on the signals that are transmitted over it, namely, the attenuations and delay. Because of the impedance discontinuities characterizing the PLC medium, the signals are reflected several times, which results in a multipath transmission, which is an effect well known in the wireless environment.

3.2.1 Channel Characterization

The powerline medium is an unstable transmission channel owing to the variance of impedance caused by the variety of appliances that could be connected to the power outlets. As these have been designed for energy distribution and not for data transmission, there are unfavorable channel characteristics with considerable noise and high attenuations. Because it is always time varying, the powerline can be considered a multipath channel that is caused by the reflections generated at the cable branches through the impedance discontinuities. The impedance of powerline channels is highly varying with frequency strongly depending on the location type and varying in a range between some few ohms up to a few kilo-ohms. The impedance is mainly influenced by the characteristic impedance of the cables, the topology of the considered part of network and the nature of the connected electrical loads. Statistical analysis of some achieved measurements has shown that nearly over the whole spectrum the mean value of the impedance is between 100 and 150 Ω. However, below 2 MHz, this mean value tends to drop toward lower values between 30 and 100 Ω. Owing to this variance of impedance, mismatched coupling in and out and the resulting transmission losses are common phenomena in the PLC networks [Phil00].

Different approaches have been proposed to describe the channel model of the powerline medium. A first approach consists of considering the PLC medium as a multipath channel, because of the multipath nature of powerline that arises from the presence of several branches and impedance mismatches that cause many signal reflections. Although this approach on which the book focuses has proven to yield a good match between the measurements and the theoretical model, as is widely investigated in [ZimmDo00a, Phil00], it has two major disadvantages. Firstly, there is a high computational cost in estimating the delay, the amplitude and the phase associated with each path. Secondly, since it is a time-domain approach, it is also necessary to take into consideration the very high number of paths associated with all the possible reflections from the unmatched terminations along the line.

Because of that, another approach has also been proposed, in which the equivalent circuits of the differential mode and the pair mode propagating along the cable are derived, and then the derived model is presented in terms of cascaded two-port networks (2PNs). Once the equivalent 2PN representation is obtained, the powerline link is represented by means of transmission matrices, also called *ABCD matrices* [BanwGa01].

3.2.2 Characteristics of PLC Transmission Cable

The propagation of signals over powerline introduces an attenuation, which increases with the length of the line and the frequency. This attenuation is a function of the powerline characteristic impedance Z_L and the propagation constant γ. According to [ZimmDo00a] and [AndrMa03], these two parameters can be defined by the primary resistance R' per unit length, the conductance G' per unit length, the inductance L' per unit length and the capacitance C' per unit length, which are generally frequency dependent, as formulated by Eqs. (3.1) and (3.2).

$$Z_L = \sqrt{\frac{R'(f) + j2\pi \cdot L'(f)}{G'(f) + j2\pi \cdot C'(f)}} \tag{3.1}$$

and

$$\gamma(f) = \sqrt{(R'(f) + j2\pi f \cdot L'(f)) \cdot (G'(f) + j2\pi f \cdot C'(f))} \tag{3.2}$$

$$\gamma(f) = \alpha(f) + j\beta(f) \tag{3.3}$$

By considering a matched transmission line, which is equivalent to regarding only the propagation of the wave from source to destination, the transfer function of a line with length l can be formulated as follows

$$H(f) = e^{-\gamma(f) \cdot l} = e^{-\alpha(f) \cdot l} \cdot e^{-j\beta(f) \cdot l} \tag{3.4}$$

In different investigations and measurements of the properties of the energy cables, it has been concluded that $R'(f) \ll 2\pi f L'(f)$ and $G'(f) \ll 2\pi f C'(f)$ in the considered frequency bandwidth for PLC (1–30 MHz). Moreover, the dependency of L' and C' on frequency is neglected so that the characteristic impedance Z_L and the propagation constant γ can be determined using the following approximations; [ZimmDo00a]:

$$Z_L = \sqrt{\frac{L'}{C'}} \tag{3.5}$$

and

$$\gamma(f) = \underbrace{\frac{1}{2} \cdot \frac{R'(f)}{Z_L} + \frac{1}{2} \cdot G'(f) Z_L}_{\text{Re}\{\gamma\}} + \underbrace{j2\pi f \sqrt{L'C'}}_{\text{Im}\{\gamma\}} \tag{3.6}$$

To get the expression for the reel part Re{} of the propagation constant as a direct function of frequency f, we substitute $R'(f)$ by its formula given in Eq. (3.7) where μ_0 and κ represent the permeability constant and the conductivity; respectively; and r is the cable radius.

$$R'(f) = \sqrt{\frac{\pi \mu_0}{\kappa r^2} f} \tag{3.7}$$

The measurements have shown that $G'(f) \sim f$, and this is also substituted into the expression of the reel part, as expressed in Eq. (3.8).

$$\alpha(f) = \text{Re}\{\gamma\} = \frac{1}{2Z_L}\sqrt{\frac{\pi\mu_0}{\kappa r^2}}f + \frac{Z_L}{2}f \qquad (3.8)$$

By summarizing the parameters of the cable (Z_L, r, etc.) into the constants k_1, k_2 and k_3, the real and the imaginary part of the propagation constant can be expressed by:

$$\alpha(f) = \text{Re}\{\gamma\} = k_1 \cdot \sqrt{f} + k_2 \cdot f \qquad (3.9)$$

$$\beta(f) = \text{Im}\{\gamma\} = k_3 \cdot f \qquad (3.10)$$

The results obtained from the diverse achieved measurements of the propagation loss were compared with the values obtained from Eq. (3.9), and an approximation was done in order to get an equation representing the real (or near the real) propagation loss behavior in frequency domain, which was presented. The approximated formulation of this loss is given by Eq. (3.11), where a_0, a_1 and k are constants.

$$\alpha(f) = a_0 + a_1 \cdot f^k \qquad (3.11)$$

Measurements of the propagation loss over the whole PLC spectrum can be found in [AndrMa03]. If the propagation loss calculated above represents the loss of the medium per unit length, then the attenuation over a medium is a function of its length l. By a suitable selection of the attenuation parameters a_0, a_1 and k, the powerline attenuation, representing the amplitude of the channel transfer function, can be defined by the Eq. (3.12) [ZimmDo00a].

$$A(f, l) = e^{-\alpha(f) \cdot l} = e^{(a_0 + a_1 \cdot f^k) \cdot l} \qquad (3.12)$$

3.2.3 Modeling of the PLC Channel

In addition to the frequency dependent attenuation that characterizes the powerline channel, deep narrowband notches occur in the transfer function, which may be spread over the whole frequency range. These notches are caused by multiple reflections at impedance discontinuities. The length of the impulses response and the number of the occurred peaks can vary considerably depending on the environment. This behavior can be described by an "echo model" of the channel as illustrated in Fig. 3.12.

Complying with the echo model, each transmitted signal reaches the receiver over N different paths. Each path i is defined by a certain delay τ_i and a certain attenuation factor C_i. The PLC channel can be described by means of a discrete-time impulse response $h(t)$ as in Eq. (3.13).

$$h(t) = \sum_{i=1}^{N} C_i \cdot \delta(t - \tau_i) \Leftrightarrow H(f) = \sum_{i=1}^{N} C_i \cdot e^{-j2\pi f \tau_i} \qquad (3.13)$$

Factoring in the formula of the channel attenuation, the transfer function in the frequency domain can be written as

$$H(f) = \sum_{i=1}^{N} g_i \cdot A(f, l_i) \cdot e^{-j2\pi f \tau_i} \qquad (3.14)$$

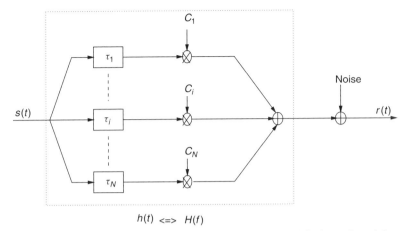

Figure 3.12 Echo model representing the multipath PLC channel model

where g_i is a weighting factor representing the product of the reflection and transmission factors along the path. The variable τ_i, representing the delay introduced by the path i, is calculated by dividing the path length l_i by the phase velocity v_p; [ZimmDo00a].

By replacing the medium attenuation $A(f, l_i)$ by the expression given in Eq. (3.12), the final equation of the PLC channel model is obtained, encompassing the parameters of its three characteristics, namely, the attenuation, impedance fluctuations and multipath effects. This equation is mainly composed of a weighting term, an attenuation term and a delay term:

$$H(f) = \sum_{i=1}^{N} \underbrace{g_i}_{\text{Weighting term}} \cdot \underbrace{e^{(a_0 + a_1 \cdot f^k) \cdot l_i}}_{\text{Attenuation term}} \cdot \underbrace{e^{-j2\pi f \tau_i}}_{\text{Delay term}} \qquad (3.15)$$

3.3 Electromagnetic Compatibility of PLC Systems

PLC technology uses the power grid for the transmission of information signals. From the electromagnetic point of view, the injection of the electrical PLC signal in the power cables results in the radiation of an electromagnetic field in the environment, where the power cables begin acting like antennas. This field is seen as a disturbance for the environment and for this reason its level must not exceed a certain limit, in order to realize the so-called *electromagnetic compatibility*. Electromagnetic compatibility means that the PLC system has to operate in an environment without disturbing the functionality of the other system existing in this environment.

In this section, after giving an exact definition of EMC, we define different aspects and terms of this concept. Then, two ways for the classification of electromagnetic disturbances are discussed. To be able to describe the real electromagnetic influence of the PLC systems on its environment, several measurements have been achieved, and the results of some of these are reported in this section. The measurements were a starting point of the standardization efforts for PLC systems for fixing the limits of the allowed electric (and also the magnetic in some cases) radiated field in their environments. Different standards, standard proposals and standardization bodies are considered in this section.

3.3.1 Different Aspects of the EMC

3.3.1.1 Definition of EMC Terms

Electromagnetic compatibility is the ability of a device or system to function satisfactorily in its electromagnetic environment without introducing intolerable electromagnetic disturbances in the form of interferences to any other system in that environment, even to itself. EMC means living in harmony with others and that has to be viewed from two aspects:

- *To function satisfactorily*, meaning that the equipment is tolerant of others. The equipment is not susceptible to electromagnetic (EM) signals that other equipment puts into the environment. This aspect of EMC is referred to as *electromagnetic susceptibility* (EMS)
- *Without producing intolerable disturbances*, meaning that the equipment does not bother other equipment. The emission of EM signals by the equipment does not cause electromagnetic interference problems in other equipment that is present. This EMC behavior is also pointed out as *electromagnetic emission* (EME)

The two mean aspects, EME and EMS, and their different variants are presented in Fig. 3.13. The concept of susceptibility is complementary to another EMC concept, which is immunity, causing, most of the time, a kind of confusion between both terms. The two terms have quite different meanings. Susceptibility is a fundamental characteristic of a piece of equipment and one can find an EM environment that will adversely affect that equipment. Immunity, on the other hand, when measured in a certain way, indicates to what extent the environment may be EM polluted before the equipment is adversely affected; [Goed95].

The electromagnetic noise propagates by conduction and by radiation, and therefore the emission can have consequences both inside and outside of the system, containing the source of the disturbances. In case of EME realization by conducted emissions, we can talk about the intrasystem compatibility; and in the case of EMC by radiated emission, the achieved compatibility is the intersystem compatibility. A similar distinction can be made for the susceptibility, where intersystem compatibility is achieved by the conducted susceptibility (CS) and the intrasystem tolerance is realized through the radiated susceptibility (RS), as presented in Fig. 3.13.

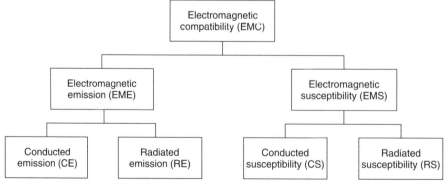

Figure 3.13 Different areas of electromagnetic compatibility

Figure 3.14 Basic model of an EMC problem

Because *electromagnetic interference* (EMI) first emerged as a serious problem in telecommunications (or, in particular, in broadcasting), EMC tends to be discussed, even to the present day, within the scope of telecommunications technology. Therefore, during the design of a telecommunications device or a system, the EMC aspect of the product must be carefully investigated before it enters the phase of wide range production. The standardization organization International Electrotechnical Commission (IEC) defined the EMI as 'degradation of the performance of a device or system by an electromagnetic disturbance'; [IEC89]. This means that the EMC problem can basically be modeled in three parts; as illustrated in Fig. 3.14:

- a source of an EM phenomenon, emitting EM energy;
- a victim susceptible to that EM energy that cannot function properly owing to the EM phenomenon; and
- a path between the source and the victim, called *coupling path*, which allows the source to interfere with the victim.

In practice, one source may simultaneously disturb several parts of equipment and several sources may also disturb a single part of equipment. However, the basic model for the investigation of EMC problems remains that in Fig. 3.14. This model allows the conclusion that if one of these three elements is absent, the interference problem is solved. For this reason, if a source of disturbance is causing many problems, it may make sense to suppress that source, that is, block the coupling path as close as possible to the source. However, not every source can be muffled up, as for example, the broadcast transmitters. A single part of an equipment that suffers interference can often be screened off, which means that the coupling path is blocked as close as possible to the affected equipment; [Goed95].

3.3.1.2 EMC Disturbance Classification

The electromagnetic disturbances from an electrical device are not easy to precisely describe, specify and analyze, but there are some general methods to classify them on the basis of some of the characteristics of the offending signals. Generally, the character, frequency content, and transmission mode provide the basis for classifying electromagnetic disturbances. A first method of classifying the EM disturbances is based on the methods of coupling the electromagnetic energy from a source to a receptor. The coupling can be in one of four categories:

- conducted (electric current),
- inductively coupled (magnetic field),
- capacitive coupled (electric field), and
- radiated (electromagnetic field).

Coupling paths often use a very complex combination of these categories making the path difficult to identify even if the source and the receptor are well known. The interference may also be radiated from the equipment via a number of different paths, depending on the frequency of that interference. For example, at high frequencies, assemblies and cables on the Printed Circuit Boards (PCBs) may strongly radiate. At lower frequencies, interference may be coupled from the equipment via the signals and the mains cables as conducted emissions. These conducted emissions may also be radiated at other different locations as further radiated emissions. Generally, the transition between radiated and conducted emissions is assumed to be around 30 MHz, where the conducted emissions dominate below this value and radiated emissions above it, as shown in Fig. 3.15.

Another way of categorizing the EM disturbances is on the basis of its three parameters: the duration, the repetition rate and the duty cycle; [Tiha95]. The disturbances can be of long or short duration. Changes of long duration are usually not included in the domain of EMC because they mainly cause alterations in the rms (root mean square) value of the mains voltage. Those with short duration last between a few seconds down to less than a microsecond. Electromagnetic disturbances with short duration can be categorized into three classes; [Tiha95]:

- *Noise*, which is a more or less permanent alteration of the voltage curve. Noise has a periodic character and its repetition rate is higher than the mains frequency. Such noise is typically generated by electric motors, welding machines, and so on. The amplitude of noise remains typically less than the peak amplitude of the mains voltage itself.
- *Impulses*, which have positive and negative peaks superimposed on the mains voltage. Impulses are characterized by having short duration, high amplitude and fast rise and/or fall times. Impulses can run synchronously or asynchronously with the mains frequency. Noises, created during various switching procedures, can exist between impulses. Typical devices that produce impulses are switches, relay controls and rectifiers.
- *Transients*, whose time period can range from a few periods of industrial frequency to a few seconds. Most commonly, transients are generated by high-power switches. To

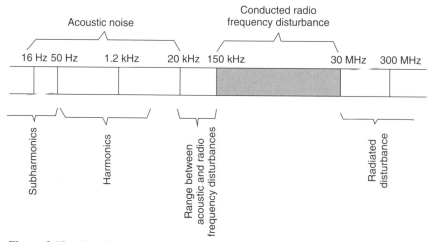

Figure 3.15 Classification of EMC disturbances according to the occupied spectrum

be able to differentiate transients from continuous noise, the duty cycle δ is introduced and defined by Eq. (3.16) [Tiha95]:

$$\delta = \tau \times f \qquad (3.16)$$

where
- τ: the pulse width measured at 50% height
- f: the pulse repetition rate, or average number of pulses per second, at random.

An electrical equipment having a duty cycle (δ) lower than 10^{-5} can be regarded as a source of transients. When the duty cycle becomes significantly higher than 10^{-5}, as with switched mode power supplies, the emitting source is no longer regarded as transient or impulse but as continuous.

To allow a systematic approach, a standard of the IEC TC 77 has established a classification of electromagnetic phenomena, which is also adopted by the European standardization CENELEC TC 210 [IEC01]. This approach is a kind of combination of both the previously discussed classification methods, as listed in Tab. 3.1.

3.3.1.3 EMI Environment Matrix

Before implementing a telecommunications system in a given location, a so-called EMI matrix has to be set. This matrix gives an idea about electromagnetic harmony between the new system and the already existing systems. A general representation of the EMI matrix of a given environment contains the elements a_{ij}, with the form presented by Eq. (3.17). The elements of the matrix can be either "+", "0" or "−". If the a_{ij} is a "+", this means that the system S_i and system S_j are tolerable and can operate simultaneously in the same location without any modifications in both systems. With a_{ij} equal to "0", a low level of EM disturbance appears in that environment and some corrections have to

Table 3.1 Principal EMC disturbances phenomena according to IEC TC 77

Low frequency		High frequency	
Conducted phenomena	Radiated phenomena	Conducted phenomena	Radiated phenomena
Harmonics, interharmonics Signaling systems Voltage fluctuations Voltage dips and interruptions Voltage unbalance Power frequency variations Induced low-frequency voltages DC in AC networks	Magnetic fields: • continuous • transient Electric fields	Directly coupled or induced voltages or currents: • continuous waves • modulated waves Undirected transients (single or repetitive) Oscillatory transients (single or repetitive)	Magnetic fields Electric fields Electromagnetic fields • continuous waves • modulated waves Transients

be done either in system i or in system j, to allow normal working for both systems. In the last case, strong corrections or radical modifications have to be effected to the new system to be able to reach normal working for both systems. In that case, it is also likely that no kind of tolerance is possible between both systems.

$$M_{\text{EMI}} = \begin{array}{c} \\ S_1 \\ S_i \\ \vdots \\ S_s \end{array} \begin{array}{cccc} S_1 & \cdots & S_j & S_s \\ \left[\begin{array}{cccc} a_{1,1} & a_{1,2} & \cdots & a_{1,s} \\ \cdots & \cdots & a_{i,j} & \cdots \\ \cdots & \cdots & \cdots & \cdots \\ a_{s,1} & a_{s,2} & \cdots & a_{s,s} \end{array} \right] \end{array}$$

In order to be able to imagine the possible sources of EM disturbances for powerline communications systems and also the possible victims of the disturbances caused by PLC equipment, Tab. 3.2 summarizes some of the already existing services and equipment operating in the frequency spectrum [1.3–30 MHz], where the broadband PLC systems are also operating. Detailed information about the complete and the exact frequency occupation of services can be found in Tab. 3.2 ([RA96]) for both, the UK's standards and the international standards.

Table 3.2 Possible EMC victims for the PLC and their band occupations

Service classes	Services	Occupied bands (MHz)
Broadcasting	Medium waves (MW) and Short waves (SW) broadcasting	1.3–1.6; 3.9–4.0; 5.9–6.0, 6.0–6.2; 7.1–7.3; 7.3–7.35; 9.4–9.5; 9.5–9.9; 13.5–13.6; 13.6–13.8; 15.1–15.6; 25.6–26.1
Maritime mobile	Tactical/strategic maritime Maritime Mobile S5.90 Distress and Safety Traffic	1.6–1.8; 2.04–2.16; 2.3–2.5; 2.62–2.65; 2.65–2.8; 3.2–3.4; 4.0–4.4; 6.2–6.5; 8.1–8.8; 12.2–13.2; 16.3–17.4; 18.7–18.9; 22.0–22.8; 25.0–25.21
	Naval broadcast communications	1.6–1.8
	Maritime DGPS	1.8–2.0; 2.0–2.02
Radio Amateur	Datamode, CW, fax, phone, etc.	1.81–1.85; 3.5–3.8; 7.0–7.1; 10.1–10.15; 14.0–14.2; 14.25–14.35; 18.0–18.16; 21.0–21.4; 24.8–24.9; 28.0–29.7
Military	NATO & UK long-distance communications	2.0–2.02; 2.02–2.04; 2.3–2.5
Aeronautical	Aeronautical	2.8–3.0; 3.02–3.15; 3.4–3.5; 3.8–3.9; 4.4–4.65; 5.4–5.68; 6.6–6.7; 8.81–8.96; 10.0–10.1; 10.1–11.1; 21.0–22.0; 23.0–23.2
Radio astronomy	Radio Astronomy	13.3–13.4; 25.55–25.67

3.3.2 PLC EM Disturbances Modeling

3.3.2.1 Source of Conducted and Radiated Disturbances

The electromagnetic emissions produced by power electronic equipments are usually broadband and coherent, occupying a wider band around the operating frequency (in megahertz range). Conducted emissions should usually be measured within this frequency range, but the standards for their measurement address these measurements only in the frequency spectrum of 0.15 to 30 MHz.

Electromagnetic disturbances can appear in the form of "common mode" (also called "asymmetrical mode") and "differential-mode (or "symmetrical mode") voltage and current. The definition of the common mode and the differential mode is shown in Fig. 3.16. The components of these modes are defined by the voltages and currents, measured on the mains terminals, and are expressed as follows; [Tiha95]:

$$U_d : U_1 - U_2$$

and

$$I_d : \frac{I_1 - I_2}{2}$$

$$U_c : \frac{U_1 + U_2}{2}$$

and

$$I_c : I_1 + I_2$$

where
- U_d = the differential-mode voltage component
- I_d = the differential-mode current component
- U_c = the common-mode voltage component
- I_c = the common-mode current component

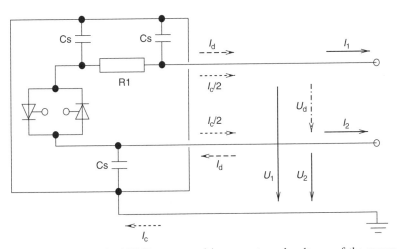

Figure 3.16 Model of a typical EMI source and its currents and voltages of the common mode and the differential mode

The general model of an EMI source is illustrated in Fig. 3.16. According to this model, a system or device that is considered an EMI source injects two types of currents into the mains network – one is in the differential mode (I_d) and the other one is in the common mode (I_c). Generally, if we inject a current signal in a cable (or wire), this one reacts as an antenna radiating an electromagnetic field into the environment, and this is also the case with the current signals I_d and I_c. The source generates a differential-mode current into the supply network in the uplink direction (from the device to mains supply), which results in the first EM field, and another differential-mode current with the same intensity as the first one in the opposite direction (from network to device). This second differential-mode current also generates an EM field with the same intensity as the field generated in the uplink direction, but in the opposite direction. As a result of symmetry, the generated EM fields wipe out each other and so no EM disturbance from the symmetrical-mode current can propagate in the environment. In the opposite to the differential-mode, the current signal in the common mode flows in the same direction. Therefore, the resulting EM fields are propagating in an asymmetrical mode, and the total field radiated in the environment is the superposition of these two fields. For this reason, the cause of the EM disturbances in the PLC networks is the absence of the common-mode disturbance.

The high-frequency (HF) equivalent circuit of an EMI source is shown in Fig. 3.17. The differential-mode current component flows in the supply wires (with the neutral wire). The differential-mode voltage component can also be measured between phase conductors. The component of the common-mode current flows from the phase and neutral conductor toward the earth. The circuit for the common-mode component is closed by the impedance Z_c. From the figure, one can conclude that there is no simple relation between the common-mode EMI components and the voltage of the EMI source, because the measured EMI depends on the mains impedance and different parasitic effects (included in Z_c), which strongly presents in the case of powerline networks.

In this high-frequency range also, the component of differential-mode current (I_d) flowing from the source to the mains networks generates an electric field, but this field is attenuated by an opposite electric field with the same strength and is generated by the current I_d flowing from the opposite side (from the network to the EMI source), as shown

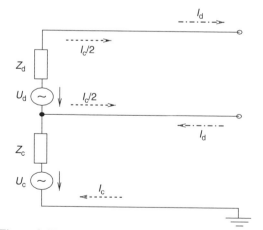

Figure 3.17 High-frequency model of an EMI source

on the HF model. Contrary to the differential mode, the current I_c of the common mode generates an electric field, without having a symmetric component that could cancel this field. From this effect comes the radiated EMI in the range 0.15 to 30 MHz.

3.3.2.2 PLC Electric Field Measurements

Normally, the electric field E is the field used for the evaluation and characterization of radio disturbances. There are problems concerning the measurements of that field. While all three components (in the three space dimension) of the magnetic field can be measured with enough precision using available sensors, the sensitivity of the available field sensors on the market permit only the measurement of the vertical field component. Owing to this lack, magnetic field loops have been used for PLC emission measurements; Fig. 3.18, and then transformed into electric fields by multiplying with the wave impedance, which is equal to the free space impedance 377 Ω. However, the calculation of the electric field components from the magnetic ones can be done only in the far field. In the near field, the wave impedance is a factor two to three times greater than the 377 Ω, which gives errors in the near field specific for PLC if it is calculated by this impedance value.

Because of the high gradient of the wave impedance variations, it is practical to take an average value for the wave impedance in the near field to be able to calculate the electric field from the measured value of the magnetic field. This means, an error estimated to be equal to factor 2 can occur by transforming the measured value of the magnetic into the electric field; [Iano02].

Because the electric field strength depends on several parameters of the powerline networks, such as the geometry, the load, and so on, and in order to give a rule for the emission field estimation, a "coupling factor" has been defined in [PLCforum]. If a mean value of this factor is defined, it could be used to determine the real field levels obtained by measuring (or knowing) the voltage or power of the injected communications signal in the power network. The coupling factor is then a function of the magnetic field and the injected energy according to the following equation:

$$k_H(\text{dB}) = 20 \log \left(\frac{H(f)}{U_{\text{inj}}(f)} \right) \tag{3.18}$$

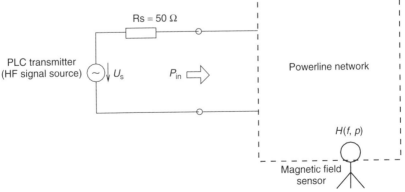

Figure 3.18 A setup for PLC radiated field measurement using a magnetic loop

Knowing the real voltage injected during the transmission over a PLC, the associated magnetic field radiated by the network can be calculated by the coupling factor. Then the radiated electric field can be easily found from the magnetic field H and the free space impedance ($Z = 377\,\Omega$) by the following equation:

$$E_{\text{PLC}}(\text{V/m}) = Z \cdot H \tag{3.19}$$

Measurements of magnetic field were conducted in different areas in order to define the coupling factor for various configurations, in-house and outdoor. The results were very similar for some installations but very different for other ones; [Iano02]. Therefore, it is not possible to use a unique coupling factor in the standardization. Different characterizing coupling factors could be defined, according to the network configuration, the environment locations, the power lines parameters, and so on.

As the radiated field from the PLC networks is caused by the asymmetrical voltage part (or the common mode) of the signal transmitted over the power lines, other investigations were proposed to directly measure this asymmetrical part of the signal and to deduce from it the strength of the radiated field; [Vick00]. In other words, it is important to determine the amount of the differential-mode signal (transverse signal) and what is converted into a common-mode signal (longitudinal signal). For this purpose, the "Longitudinal Conversion Loss" (LCL) and the "Transversal Conversion Loss" (TCL) methods were first defined in the ITU recommendations for all types of networks, before being adopted in the ETSI standards definitions and measurements set up for the PLC, in the report titled "Power Line Telecommunications (PLT) Channel Characterization and Measurement Methods" [ETSI03].

The LCL and TCL are ratios between the asymmetric and the symmetric components of the voltage at a specific test point in the PLC network. The LCL of a specific test point is determined by coupling an asymmetrical voltage (or longitudinal signal) into the system and measuring the resulting symmetrical voltage (or transversal signal). The LCL is a logarithmic ratio between the asymmetrical component (E_L) and the resulting symmetrical voltage (V_T) according to the following relation:

$$LCL(\text{dB}) = 20\log\left(\frac{E_L}{V_T}\right) \tag{3.20}$$

The TCL is the ratio between the symmetrical and the asymmetrical voltage when a symmetrical voltage is injected into the transmission line.

$$TCL(\text{dB}) = 20\log\left(\frac{E_T}{V_L}\right) \tag{3.21}$$

These methods can be applied to all telecommunications systems, such as transmission lines, equipment or their combinations. However, the TCL is the most important value with respect to being able to determine the amount of the longitudinal (or common mode) voltage caused by unbalances in the system, which is the principle cause of the radiated disturbances. Once the TCL is known, one would be able to calculate the asymmetric voltage at a given amplitude of the symmetrical signals. Then, this can be used to estimate the strength of the radiated emissions with an appropriate model.

3.3.3 EMC Standards for PLC Systems

3.3.3.1 EMC Standardization Organizations

EMC standards are prerequisites to insure that the numerous devices and systems do not disturb each other or give rise to malfunctioning of some of them. They lay down requirements for equipment as regards both the maximum permitted emission of parasitic conducted and radiated electromagnetic disturbances, as well as the availability of the equipment under the influence of these disturbances. To test the equipment and to check if it respects the emission limits, test setups to measure the disturbance levels are also defined by the standards. However, standards are only one aspect of the problems associated with the EMC.

The EMC standardization bodies are categorized in three classes, according to the number of states in which they operate: international; regional, the most representative of which are those of the United States and the European Union; and national, such as RegTP in Germany and RA (Radiocommunications Agency) in the United Kingdom. All these bodies work in a consultative and cooperative way to develop EMC standards, which try to combine the interest of all parts whose relationship is shown in Fig. 3.19.

International Committees

The International Electrotechnical Commission is an organization that promotes and coordinates international standardization and related matters, such as the assessment of conformity to standards, in the fields of electricity, electronics and related technologies. For its technical work, the IEC comprises some 200 committees and subcommittees, of which about 50 are concerned with EMC in varying degrees. These committees and subcommittees present the results of their work in the form of standards or technical reports. The oldest and most important one of these committees is the "Comité International Spécial des Perturbations Radio Electriques" or international committee for radio interferences (CISPR), which was set up by the IEC in 1934 in Paris, when radio frequency interferences (RFI) had begun to be a problem. This was the first international coordinating organization to produce standards to protect the reception of radio transmission and has extended its field of activity to EMC product standards, for example, for household equipment and Information Technology Equipment (ITE). Its recommendations contained

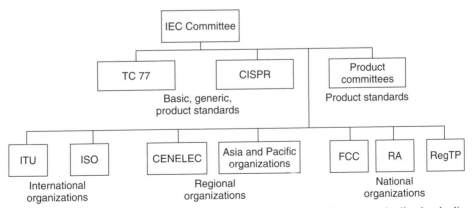

Figure 3.19 Organization of EMC work and liaisons between different standardization bodies

in CISPR 22 had defined limits for the conducted and radiated emissions from ITEs and served as the basis for the major national standards.

The second important standardization subcommittee of the IEC is the Technical Committee 77 (TC 77), also referred to as IEC TC 77 and it plays a complementary role to the CISPR. It was created in 1973, to be responsible, together with other committees to some extent, for Basic EMC standards that have general application and for Generic EMC standards, in which the stated requirements can be fully or partially respected; [IEC01]. It also allows a systematic approach for classifying the EM phenomena. The study of the EMS of electrical equipment and articulation of measurement methods, as well as the compilation of recommendations and standards for this domain of EMC, has been the specialty of IEC subcommittee TC 65.

Regional Organizations
In the United States, the Federal Communications Commission (FCC) is the governmental agency that is responsible for the frequency planning and interference control. Most of the time, the FCC is considered a regional organization rather than a national organization. The FCC has regulations covering the limitation of emissions from a wide range of products; among these the FCC Part 15 standards are applied to all digital equipment. These FCC regulations impose two different emission limits, measured at different distances from the device. The applicable limit depends on the environment in which the equipment will operate. Class A equipment is designed for use in commercial or industrial applications. Class B defines limits to be applied to equipment for use at home or in residential premises. The FCC does not specify the shielding effectiveness but regulates the EM emissions for both classes A and B. For each class, FCC defines the limits of the radiated field strength in the spectrum 30 MHz to 1 GHz and the voltage limit of the conducted disturbances in the frequency band 450 kHz to 30 MHz.

In Europe and in the framework of the "Comité de Coordination Européen des Normes Electriques pour le Marché Commun" (CENELCOM), or European coordination committee of electrical standards in the European Common Market, a decision was taken to establish a Common Standardization Committee for creating a standard for electrical equipment emission limits. The Common Standardization Committee was founded in 1970 and immediately linked itself with representatives of electrical energy suppliers and electrical household appliances manufacturers. The Common European Market was enlarged in 1973 and with it CENELCOM was reorganized under the name CENELEC, for "Comité Européen de Normalization Electrotechnique", or European electrical standardization committee. There are series of European EMC standards for various types of specific equipment, such as information technology, but there are also general emission (EN 50081) and immunity standards, which apply in the absence of specific standards; [Moly97].

National Regulators
The Radiocommunications Agency is an Executive Agency of the UK's Department of Trade and Industry. It is responsible for the management of the nonmilitary radio spectrum in the United Kingdom, which involves international representation, commissioning research, allocating spectrum and licensing its use, and keeping the radio spectrum clean; [Stro01]. In Germany and after the liberalization of the postal and telecommunications markets, the

former monopoly operators, Deutsche Post AG and Deutsche Telekom AG were still maintaining dominant positions in the market. From here came the need for a regulatory body, which had to keep a check on each dominant provider in order to create a level playing field to protect the new entrants. A structurally separate authority with maximal possible independence was needed to perform this task. The "Regulierungsbehörde für Telekommunikation und Post" (RegTP), or the regulatory authority for telecommunications and posts was therefore set up on 1 August 1996. It is equipped with effective procedures and instruments with which to enforce the regulatory aim. These include information and investigative rights as well as a set of sanctions.

3.3.3.2 Standards for PLC Radiated Emission Limits

Under the observation of the radio communications agency in London, several field trials were first monitored to explore the PLC technology and to get an idea about the EMI caused by its equipment. In parallel, the agency called for the development of a new measured procedure called MPT1570, which was titled "electromagnetic radiations from telecommunications systems operating over material substances in the frequency range 9 kHz to 300 MHz". Measurements were led in peak mode using a magnetic loop, according to the measurement set up in Fig. 3.18, and applying the limits for the electrical field strength expressed by the equation; [Hans00]:

$$E = 20 \left(\frac{dB \, \mu V}{m} \right) - 7.7 \log \left(\frac{f}{MHz} \right) \tag{3.22}$$

Because of these low limits, the measured EMI levels from PLC systems were largely above the recommended values of the allowable electrical field strength. With these enormous approval difficulties in the United Kingdom and the massive protests of civil and military frequency users in the shortwave range, the PLC activities were shifted outside the United Kingdom especially to Germany. The German regulatory authority RegTP, under the ministry of economic affairs, published its first EM limitations in a draft paper in January 1999. This was later known under the name "NB30" and is less than the UK proposal by approximately 20 dB. These limitations, whose 3-m limits are shown in Tab. 3.3, concern not only the PLC, but every kind of wire-bound data transmission, including cable TV, xDSL, and so on; [Hans00]. The measurement setup follows the standard RegTP 322 MV 05 [RegTP].

In the United States, FCC Part 15 specifically excludes current carrier systems that are unintentional radiators, including the PLC, from conducted emissions limits above

Table 3.3 E field strength limits allowed by the NB30 for PLC and other wired systems

Frequency bands	Limits for the E field strength (peak)
0.009 MHz–1 MHz	40 dB(μV/m) $- 20 \log_{10}$ (f/MHz)
1 MHz–30 MHz	40 dB(μV/m) $- 8.8 \log_{10}$ (f/MHz)
30 MHz–1 GHz	27 dB(μV/m)
1 GHZ–3 GHz	40 dB(μV/m)

1.705 MHz, relying instead on specified radiated emissions limits. The allowed limits for the radiated E field according to the FCC Parts are shown in Tab. 3.4, as it was published in the 10/01/1999 edition; [FentBr01].

The American standard offers a wide horizon for the implementation of the powerline communications, with its high tolerance for the delivered EMI. Investigations were achieved about the distribution of the channel capacity when the radiation complies with FCC Part 15, and NB30. This channel investigation shows that the capacity of power line channel is larger than 63 Mbps when the FCC mask is used, and larger than 3 Mbps when the NB30 limits are used; [EsmaKs02].

For a qualitative comparison between the three standards, MPT1570, NB30 and FCC parts 15, Fig. 3.20 was elaborated showing the limits for the radiated E field in the spectrum 1–30 MHz. For this purpose, FCC Part 15 limits are extrapolated from a 30- to a 3-m measurement distance using a factor of 20 dB per decade. This extrapolation is valid only for the purpose of comparing the emission limits, and actual measurements can be achieved at measurement distances other than the 30 m, whilst different extrapolation factors can also be applied; [FentBr01].

During 2001, the European Commission issued a mandate, M313, and invited people from different standard bodies including CENELEC, ETSI, CISPR and members from

Table 3.4 Recommended E field strength for PLC systems according to FCC part 15

Frequency band (MHz)	Radiated emission limit (μV/m) (peak)	Measured at (m)
1–1.705	15	47,715/frequency (kHz)
1.705–10	100	30
10–13.553	30	30
13.553–13.567	10,000	30
13.567–26.96	30	30
26.96–27.28	10,000 (average)	3
27.28	30	30

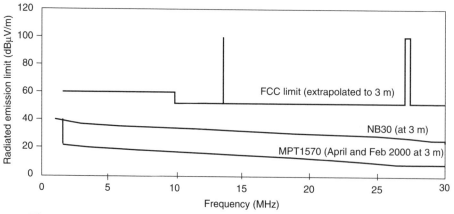

Figure 3.20 Radiated emission limits from MPT1570, NB30 and FCC Part 15

CEPT (Conference of European Post and Telecommunications) together with PLC design houses to start the establishment of harmonized standards for all telecommunications networks. The Mandate M313 is officially titled by "Standardization Mandate Addressed to CEN, CENELEC and ETSI concerning Electromagnetic Compatibility in Telecommunications Networks". The main thrust for this mandate is the establishment of harmonized standards, considering both emission and immunity, for powerline communications systems, coaxial cables and telephone lines. The emphasis is on the communication network and not on the equipment, although the latter should be in line with any standards being produced for the EMC of equipment. Furthermore, the European commission M313 is proposed to solve the range of different emission standards into one single standard for all wired communications networks, by trying to find a kind of compromise between all possible proposals and/or standards from different bodies and countries, such as Norwegian proposal, NB30, FCC, MPT1570, and so on; [NewbYa03].

3.3.3.3 Limits for the Conducted Emissions

The CISPR 22 standard presents procedures for the measurement of the levels of the conducted emission signal generated by the Information Technology Equipment and specifies its limits for the frequency range from 0.15 MHz to 1 GHz. Like all other FCC standards, this standard also subdivides the equipment into two categories: class A and class B. Different limits are applied to these classes, where the limits of class B are stricter than those of class A. If the tested equipment respects the limits of class A, but not those of class B, this device can be used legally if a notice is included, which indicates that this product may cause EMI.

Different limitations and measurement setups are defined for the conducted emissions from mains ports and telecommunications ports. In the actual version of the standard, the telecommunications ports are seen as ports that are intended to be connected to typical communications networks. The mains plug of a PLC equipment combines the functionality of a mains port with that of a telecommunications port. Therefore, some proposals for future amendments of the current standard include the definition of the so-called multipurpose ports and their proper measurement procedure. For the moment, the mains plug of PLC equipment falls into the category of mains port and has to be measured accordingly. The recommended limits for the conducted disturbances at mains port and telecommunications ports are given in Tabs. 3.5 and 3.6; [Hens02].

Table 3.5 Limits for conducted disturbances at the mains ports of class A and class B ITE

Frequency band (MHz)	Limits in dB(μV)			
	Class A		Class B	
	Quasi-peak	Average	Quasi-peak	Average
0.15–0.50	79	66	66–56	56–46
0.50–5	73	60	56	46
5–30	73	60	60	50

Table 3.6 Limits for conducted common-mode disturbances at telecommunications ports of class A and class B equipment

Frequency band (MHz)	Limits in dB(μV)			
	Class A		Class B	
	Quasi-peak	Average	Quasi-peak	Average
0.15–0.50	97–87	84–74	84–74	74–64
5–30	87	74	74	64

3.4 Disturbance Characterization

3.4.1 Noise Description

Because the power cables were designed only for energy transmission, no interest has been shown in the properties of this medium in the high-frequency range. Furthermore, a wide variety of appliances, with different properties, are connected to the power network. Therefore, before using this medium for information transmission, an intensive investigation of the phenomena present in their environment has to be achieved. Besides the distortion of the information signal, owing to cable losses and multipath propagation, noise superposed on the utile signal energy make correct reception of information more difficult. Unlike the other telecommunications channels, the powerline channel does not represent an Additive White Gaussian Noise (AWGN), whose power spectral density is constant over the whole transmission spectrum.

A lot of investigations and measurements were achieved in order to give a detailed description of the noise characteristics in a PLC environment. An interesting description is given in [ZimmDo00a], which classifies the noise as a superposition of five noise types, distinguished by their origin, time duration, spectrum occupancy and intensity; the approximative representation of spectrum occupation is illustrated in Fig. 3.21:

- *Colored background noise (type 1)*, whose power spectral density (psd) is relatively lower and decreases with frequency. This type of noise is mainly caused by a superposition of numerous noise sources of lower intensity. Contrary to the white noise, which is a random noise having a continuous and uniform spectral density that is substantially independent of the frequency over the specified frequency range, the colored background noise shows strong dependency on the considered frequency. The parameters of this noise vary over time in terms of minutes and hours.

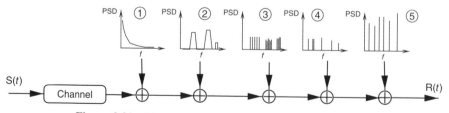

Figure 3.21 The additive noise types in PLC environments

- *Narrowband noise (type 2)*, which most of the time has a sinusoidal form, with modulated amplitudes. This type occupies several subbands, which are relatively small and continuous over the frequency spectrum. This noise is mainly caused by the ingress of broadcast stations over medium- and shortwave broadcast bands. Their amplitude generally varies over the daytime, becoming higher by night when the reflection properties of the atmosphere become stronger.
- *Periodic impulsive noise, asynchronous to the main frequency (type 3)*, with a form of impulses that usually has a repetition rate between 50 and 200 kHz, and which results in the spectrum with discrete lines with frequency spacing according to the repetition rate. This type of noise is mostly caused by switching power supplies. A power supply is a buffer circuit that is placed between an incompatible source and load in order to make them compatible. Because of its high repetition rate, this noise occupies frequencies that are too close to each other, and builds therefore frequency bundles that are usually approximated by narrow bands.
- *Periodic impulsive noise, synchronous to the main frequency (type 4)*, is impulses with a repetition rate of 50 or 100 Hz and are synchronous with the main powerline frequency. Such impulses have a short duration, in the order of microseconds, and have a power spectral density that decreases with the frequency. This type of noise is generally caused by power supply operating synchronously with the main frequency, such as the power converters connected to the mains supply.
- *Asynchronous impulsive noise (type 5)*, whose impulses are mainly caused by switching transients in the networks. These impulses have durations of some microseconds up to a few milliseconds with an arbitrary interarrival time. Their power spectral density can reach values of more than 50 dB above the level of the background noise, making them the principal cause of error occurrences in the digital communication over PLC networks.

The achieved measurements have generally shown that noise types 1, 2 and 3 remain usually stationary over relatively longer periods, of seconds, minutes and sometimes even of some hours. Therefore, all these three can be summarized in one noise class, that is seen as colored PLC background noise class and is called "Generalized background noise", whose frequency occupation and mathematical model are discussed below. The noise types 4 and 5 are, on the contrary, varying in time span of milliseconds and microseconds, and can be gathered in one noise class called "impulsive noise", pointed out also in other literatures as "impulse noise". Because of its relatively higher amplitudes, impulse noise is considered the main cause of burst error occurrence in data transmitted over the high frequencies of the PLC medium.

3.4.2 Generalized Background Noise

For the modeling of the generalized background noise in the PLC environment, it is considered as the superposition of the colored background noise and the narrowband disturbances; as illustrated in Fig. 3.22. In this case, no difference is made between the shortwave radios and the other narrowband disturbances in the form of spectral lines, because normally the spectral lines are found in bundled form. For the modeling, these bundles of disturbers are approximated by their envelope. Furthermore, because of the high repetition rate noise type (3) occupies frequencies that are too close to each other,

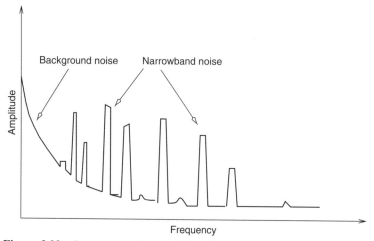

Figure 3.22 Spectral density model for the generalized background noise

and build therefore frequency bundles that are usually approximated by a narrowband occupation. Therefore, for its modeling, this noise will be seen as a narrowband noise with very low psd. The power density of the colored background noise is time-averaged for the modeling by $N_{\mathrm{CBN}}(f)$. The time-dependence characteristic of this noise can be modeled independently with the knowledge of the standard deviation; [Beny03]. Therefore, the psd of the generalized background noise can be written under the following form:

$$N_{\mathrm{GBN}}(f) = N_{\mathrm{CBN}}(f) + N_{\mathrm{NN}}(f) \tag{3.23}$$

$$N_{\mathrm{GBN}}(f) = N_{\mathrm{CBN}}(f) + \sum_{k=1}^{B} N_{\mathrm{NN}}^{(k)}(f) \tag{3.24}$$

where $N_{\mathrm{CBN}}(f)$ is the psd of the colored background noise, $N_{\mathrm{NN}}(f)$ the psd of the narrowband noise and $N_{\mathrm{NN}}^{k}(f)$ is the psd of the subcomponent k generated by the interferer k of the narrowband noise.

For the model of the colored background noise psd, the measurements have shown that a first-order exponential function is more adequate, as formulated by Eq. (3.25); [Beny03].

$$N_{\mathrm{CBN}}(f) = N_0 + N_1 \cdot e^{-\frac{f}{f_1}} \tag{3.25}$$

with N_0 the constant noise density, N_1 and f_1 are the parameters of the exponential function, and the unit of the psd is $\mathrm{dB\mu V/Hz^{1/2}}$. Through different investigations and measurements of noise in residential and industrial environments, it was possible to find out approximations for the parameters of this model and the psd of the colored background noise can be described by Eqs. (3.26) and (3.27) for residential and industrial environments respectively; [Phil00]:

$$N_{\mathrm{BN}}(f) = -35 + 35 \cdot e^{-\frac{f[\mathrm{MHz}]}{3,6}} \quad \text{for residential environments and} \tag{3.26}$$

$$N_{\mathrm{BN}}(f) = -33 + 40 \cdot e^{-\frac{f[\mathrm{MHz}]}{8,6}} \quad \text{for industrial environments} \tag{3.27}$$

For the approximation of the narrowband noise interferers, the parametric Gaussian function is used, whose main advantages are the few parameters required for specifying the model. Furthermore, the parameters can be individually found out from the measurements, which have shown only a small variance; [Beny03]:

$$N_{NN}^{(k)}(f) = A_k \cdot e^{-\frac{(f-f_{0,k})^2}{2 \cdot B_k^2}} \tag{3.28}$$

the function parameters are A_k for the amplitude, $f_{0,k}$ is the center frequency and B_k is the bandwidth of the Gaussian function.

3.4.3 Impulsive Noise

The impulsive noise class is composed of the periodic impulses that are synchronous with the main frequency and the asynchronous impulsive noise. The measurements show that this class is largely dominated by the last noise type (type 5). For this reason, the modeling of this class is based on the investigations and the measurements of type (5), of which an example is shown in Fig. 3.23.

The aim of these investigations and measurements is to find out the statistical characteristics of the noise parameters, such as the probability distribution of the impulses width and their interarrival time distribution, representing the time between two successive impulses, Fig. 3.24. One approach to model these impulses is a pulse train with pulse width t_w, pulse amplitude A, interarrival time t_a and a generalized pulse function $p(t/t_w)$ with unit amplitude and impulse width t_w; [ZimmDo00a]:

$$n_{imp}(t) = \sum_{i=-\infty}^{\infty} A_i \cdot p\left(\frac{t - t_{a,i}}{t_{w,i}}\right) \tag{3.29}$$

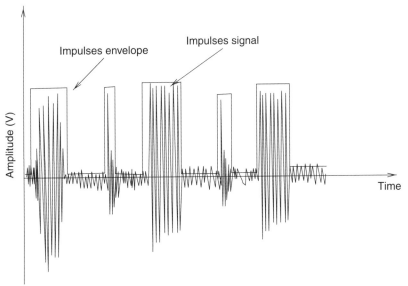

Figure 3.23 Example of some measured impulses in the time domain in a PLC network

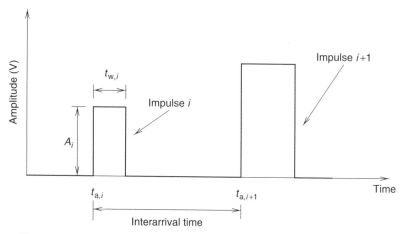

Figure 3.24 The impulse model used for impulsive noise class modeling

The parameters $t_{w,i}$, A_i and $t_{a,i}$ of impulse i are random variables, whose statistical properties are measured and investigated in [ZimmDo00a]. The measured impulses have shown that 90% of their amplitudes are between 100 and 200 mV. Only less than 1% exceeds a maximum amplitude of 2 V. The measurements of the impulse width t_w have also shown that only about 1% of the measured impulses have a width exceeding 500 μs and only 0.2% of them exceeded 1 ms. Finally, the interarrival time that separates two successive impulses is below 200 ms for more than 90% of the recorded impulses. Other more detailed measurements show that about 30% of the detected pulses had an interarrival time of 10 or 20 ms, which represents the impulsive noise that is synchronous with the mains supply frequency, noise type 3. The interarrival times, lying above 200 ms, have an exponential distribution.

3.4.4 Disturbance Modeling

The disturbances can have a big impact on the transmission in PLC networks on different network layers. As this book focuses on the design of the MAC layer, we consider the disturbance modeling to be used in such investigations. In the following section, we describe a simple on–off disturbance model and a complex disturbance model for application in investigations of OFDM-based transmission systems.

3.4.4.1 On–Off Model

In Sec. 3.4.2, it is shown that the generalized background noise is stationary over seconds, minutes or even hours. It is also concluded that periodic impulses, synchronous to the mean frequency (noise type 4) have a short duration and low psd. On the other hand, the short-term variance in the powerline noise environment is mostly introduced by the asynchronous impulsive noise (type 5). Those impulses can reach a duration of up to several milliseconds and a higher psd.

Suitable methods for forward error correction and interleaving (Sec. 4.3) can deal with disturbances caused by the impulsive noise. However, a certain error probability remains,

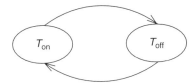

Figure 3.25 On–Off disturbance model

which results in erroneous data transmission and the resulting retransmission of the damaged data units. Incorrect data transmission has a big influence on the performance of MAC and higher network layers. Therefore, an on–off disturbance model is developed to represent the influence of the asynchronous impulsive noise on the data transmission. The noise impulses can make a transmission channel for a certain time period. After the impulse disappears, the affected transmission channel is again available. Under this kind of noise, the disturbances in a PLC transmission channel can be represented by an on–off model with two states; T_{on} and T_{off} (Fig. 3.25) [HrasHa00].

T_{off} state represents the duration of an impulse making the channel unavailable for the time of its duration. T_{on} is the time without disturbances (absence of disturbance impulses) when the channel is considered available. Both duration of the disturbance impulses and their interarrival time can be represented by two random variables that are negative exponentially distributed, according to the behavior of the noise impulses [ZimmDo00, ZimmDo00a, Zimm00].

3.4.4.2 Complex Disturbance Models for OFDM-based Systems

In the consideration above, an on–off error model is defined describing the availability of a transmission channel. However, if a disturbance impulse occurs, it can affect a variable number of OFDM subcarrier frequencies depending on its characteristics, spectral power, origin, and so on. Therefore, the disturbances have to be modeled not only in the time domain (duration and interarrival time of impulses) but also in the frequency domain, specifying how many and which subcarriers are affected by a disturbance impulse.

Furthermore, in the simple on–off disturbance model, an OFDM subcarrier can be only in two hard defined states: On – available for the transmission, or Off – not available. On the other hand, an OFDM system can apply bit loading (Sec. 4.2.1) to provide variable data rates of a subcarrier according to its quality, which depends on the noise behavior on the subcarrier frequency. To model an OFDM system using bit loading, the on–off disturbance model is extended to include several states between "channel is Off" (transmission not possible) and "channel is On" (full data rate is possible) as is presented in Tab. 3.7.

The states between "Off" and "On" represent the situations when a subcarrier is affected by the disturbance impulse, but is still able to transmit the data. In such cases, the OFDM-based systems are able to reduce the data rate over affected subcarriers and to make the

Table 3.7 Subcarrier data rates in a multistate error model – an example

Subcarrier status	On	On_{-1}	On_{-2}	On_{-3}	On_{-4}	On_{-5}	On_{-6}	On_{-7}	Off
Data rate/kbps	8	7	6	5	4	3	2	1	0

transmission possible. Therefore, the multistate error model make sense if an OFDM-based PLC system is investigated. As is mentioned above, the length of typical PLC access networks is up to several hundreds meters. Thus, we can expect that the distrubances can differently affect particular network segments; for example, depending on the position of noise source, protection of powerline grids in different network sections, and so on. In this case, a PLC network is under the influence of so-called selective distrubances, where the network stations are differently affected by particular disturbances, which primarly depend on their position in the network. Such distrubances are represented by selective disturbance models. It can be concluded that the distrubances can act selectively in two different ways, frequency and space/position dependent.

3.4.4.3 Model Parameters

For the specification of the parameters representing general disturbance characteristics in PLC access networks, measurements of the disturbance behaviors have to be carried out in numerous networks operating in various environments: rural and urban areas, business and industrial areas, PLC networks designed with various technologies (e.g. different types of cables), and so on. Local conditions and realizations of PLC networks can be very different from each other and the achieved measurement results can strongly vary from network to network. Therefore, there is not only a need for the general characterization of the disturbance behavior but also for the characterization of each individual PLC access network.

3.5 Summary

The low-voltage networks have complex topologies that can differ strongly from one network to another. This difference comes from the fact that they have parameters whose values can be varied, such as the users density, the users activity, the connected appliances, and so on. Generally, it can be concluded that low-voltage power supply networks, also including in-home part of the network, have a physical tree topology. However, on the logical level, a PLC access network can be considered a bus network, representing a shared transmission medium. Because PLC networks perform on shared medium, there is the need for medium access management policy. This task is taken by a base station, which control the access to the medium over the whole or only a part of the considered PLC network. The base station is also the point over which access to the WAN is possible. Additional PLC devices, such as repeaters and/or gateways can also be implemented.

Low-voltage networks were designed only for energy distribution to households and a wide range of devices and appliances are either switched on or off at any location and at any time. This variation in the network charge leads to strong fluctuation of the medium impedance. These impedance fluctuations and discontinuity lead to multipath behavior of the PLC channel, making its utilization for the information transmission more delicate. Beside these channel impairments, the noise present in the PLC environment makes the reception of error-free communication signal more difficult. The noise in PLC networks is diverse and is described as the superposition of five additive noise types, that are categorized into two main classes – on the one hand is the background noise, which remains stationary over long time intervals, and on the other is the impulsive noise, which consists of the principle obstacle for a free data transmission, because of its relative high intensity. This impulsive noise results in error bursts, whose duration can exceed the limit to be detected and corrected usually by used error correcting codes. Therefore, the impulsive noise in PLC networks has to be represented in appropriate disturbance models.

EMC is the first requirement to be met by any device, before it enters the market and even before it enters the wide production phase. However, this remains the main challenge that the PLC community is facing. Several services use one or multiple parts of the spectrum 0–30 MHz that is targeted by the PLC system. This makes the set of possible EM victims of PLC devices larger. In spite of it, standardization activities are going on and trying to reach international flexible standards for the electrical field strength limits, like those imposed by the FCC Part 15.

4

Realization of PLC Access Systems

As considered in Chapter 3, PLC access networks are characterized by given topology of low-voltage supply networks, unfavorable transmission conditions over power grids, problem of electromagnetic compatibility and resulting low data rates and sensitivity to disturbances from the network itself and from the network environment. To solve these problems and to be able to ensure data transmission over power grids, achieving certain data rates necessary for realization of the broadband access, various transmission mechanisms and protocols can be applied. As mentioned in Sec. 2.3.3, PLC access systems are realized by several network elements. Basically, the communication within a PLC access network takes place between a base station and a number of PLC modems, connecting PLC subscribers and their communications devices. In this chapter, we present realization of PLC access systems including their transmission and protocol architecture implemented within the network elements, as well as telecommunications services which are applied to broadband PLC networks.

4.1 Architecture of the PLC Systems

Exchange of information between distant communicating partners seems to be very complex. The communications devices used can differ from each other, and the information flow between them can be carried out over multiple networks, which can apply different transmission technologies. To understand the complex communications structures, the entire communications process has been universally standardized and organized in individual hierarchical communications layers [Walke99]. The hierarchical model exactly specifies tasks of each communications layer as well as interfaces between them, ensuring an easier specification and standardization of communications protocols.

Nowadays, the ISO/OSI Reference Model (*International Standardization Organization/Open Systems Interconnection*, Fig. 4.1) is mainly used for description of various communications systems. It consists of seven layers, each of them carrying a precisely defined function (or several functions). Every higher layer represents a new level of abstraction compared to the layer below it. The first network layer specifies data transmission on a so-called physical network layer (transmission medium), and every higher

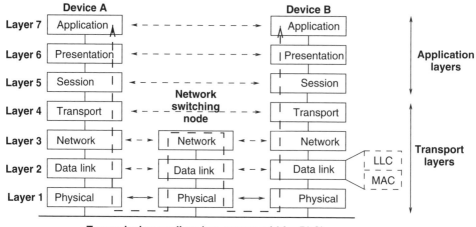

Figure 4.1 The ISO/OSI reference model

layer specifies processes nearer to communications applications (end user device). The OSI reference model is well described in the available literature, for example, [Tane98]. Therefore, we just give a brief description of functions specified in the reference model so as to be able to define PLC specific network layers.

- Layer 1 – Physical Layer – considers transmission of bits over a communications medium, including electrical and mechanical characteristics of a transmission medium, synchronization, signal coding, modulation, and so on.
- Layer 2 – Data Link – is divided into two sublayers (e.g. [John90]):
  - MAC – Medium Access Control (lower sublayer) – specifies access protocols
  - LLC – Logical Link Control (upper sublayer) – considers error detection and correction, and data flow control.
- Layer 3 – Network Layer – is responsible for the set-up and termination of network connections, as well as routing.
- Layer 4 – Transport Layer – considers end-to-end data transport including segmentation of transmitted messages, data flow control, error handling, data security, and so on.
- Layer 5 – Session Layer – controls communication between participating terminals (devices).
- Layer 6 – Presentation Layer – transforms data structures into a standard format for transmission.
- Layer 7 – Application Layer – provides interface to the end user.

Network layers 5–7 are nearer to the end user and to a running communications application. Therefore, these network layers are very often characterized as *Application Network Layers* (or Application-oriented Layers) [Kade91]. As against the application layers, network layers 1–4 are responsible for the transmission over a network, and accordingly, they are called *Transport Layers* (Fig. 4.1), or *Transport-oriented Layers*.

As mentioned above, the transport layer (layer 4) takes care of end-to-end connections and, accordingly, is implemented within end communication devices (e.g. TCP in standard computer equipment). On the other hand, network layers 1–3 fulfill tasks related to the data transmission over different communications networks and network sections (subnetworks). In accordance with this, these layers are implemented within various network elements, such as switching nodes, routers, and so on, and are called *Network Dependent Layers* (or Network Layers). Thus, the transport layer (layer 4) represents an interface between the network layers and the totally network-independent application layers 5–7.

A PLC access network consists of a base station and a number of subscribers using PLC modems. The modems provide, usually, various user interfaces to be able to connect different communications devices (Fig. 4.2). Thus, an user interface can provide an Ethernet interface connecting a personal computer. On the other hand, a PLC modem is connected to the powerline transmission medium providing a PLC specific interface. The communication between the PLC transmission medium and the user interface is carried out on the third network layer. Information received on the physical layer form the powerline network is delivered through MAC and LLC sublayers to the network layer, which is organized according to a specified standard (e.g. IP) ensuring communications between PLC and Ethernet (or any other) data interfaces. The information received by the data interface of the communications device is forwarded to the application network layers.

The base station connects a PLC access network and its powerline transmission medium to a communications distribution network, and with it to the backbone network (Sec. 2.3.4). Accordingly, it provides a PLC specific interface and a corresponding interface to the communications technology used in the distribution network. Generally, the data exchange between a PLC network and a distribution network is carried out on the third network layer, such as between the PLC interface in the modem and the user interface.

In accordance with the consideration presented above, it can be recognized that both base stations and PLC modems provide a specific interface for their connection to the powerline transmission medium (Fig. 4.2). On the other hand, the interfaces for the connection to the distribution and backbone networks, as well as to various communications devices, are realized according to communications technologies applied in the backbone and in the end devices, which are specified in the corresponding telecommunications standards. The interconnection between PLC and other communications technologies is carried out on the third network layer, which is also standardized.

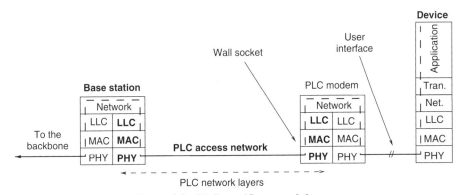

Figure 4.2 PLC specific network layers

The PLC specific interface includes first two network layers: physical layer and MAC and LLC sublayers of the second network layer. PLC physical layer is organized according to the specific features of the powerline transmission medium and is described in Sec. 4.2. Owing to the inconvenient noise scenario in PLC networks (Sec. 3.4), various mechanisms for error handling, as a part of the LLC sublayer, are an important issue and they are considered in Sec. 4.3. A description of PLC services and their classification are presented in Sec. 4.4. Because of the fact that the emphasis of this book is set on the MAC sublayer, PLC MAC layer and its protocols are separately considered in Chapter 5 and Chapter 6.

4.2 Modulation Techniques for PLC Systems

The choice of the modulation technique for a given communications system strongly depends on the nature and the characteristics of the medium on which it has to operate. The powerline channel presents hostile properties for communications signal transmission, such as noise, multipath, strong channel selectivity. Besides the low realization costs, the modulation to be applied for a PLC system must also overcome these channel impairments. For example, the modulation, to be a candidate for implementation in PLC system, must be able to overcome the nonlinear channel characteristics. This channel nonlinearity would make the demodulator very complex and very expensive, if not impossible, for data rates above 10 Mbps with single-carrier modulation. Therefore, the PLC modulation must overcome this problem without the need for a highly complicated equalization. Impedance mismatch on power lines results in echo signal causing delay spread, consisting in another challenge for the modulation technique, which must overcome this multipath. The chosen modulation must offer a high flexibility in using and/or avoiding some given frequencies if these are strongly disturbed or are allocated to another service and therefore forbidden to be used for PLC signals.

Recent investigations have focused on two modulation techniques that have shown good performances in other difficult environment and were therefore adopted for different systems with wide deployment. First, the Orthogonal Frequency Division Multiplexing (OFDM), which has been adopted for the European Digital Audio Broadcasting (DAB), the Digital Subscriber Line (DSL) technology, and so on. Second, the spread-spectrum modulation, which is widely used in wireless applications, offering an adequate modulation to be applied with a wide range of the multiple access schemes.

In this section, we explain the principles of each modulation technique and their mathematical background. Then, some practical realizations of the demodulator (or transmitter) and its corresponding demodulator (or receiver) are proposed for each modulation. Finally, a comparison between these candidates is discussed, showing the advantages and drawbacks of each one of them. This comparison could make it possible to make a decision about the choice of the modulation technique to be adopted for PLC systems, allowing to meet some performances that can be required from the network, such as the high bit rate, the level of electromagnetic disturbances, or bit error rate, and so on.

4.2.1 Orthogonal Frequency Division Multiplexing

4.2.1.1 Modulation Principles

MultiCarrier Modulation (MCM) is the principle of transmitting data by dividing the stream into several parallel bit streams, each of which has a much lower bit rate, and by

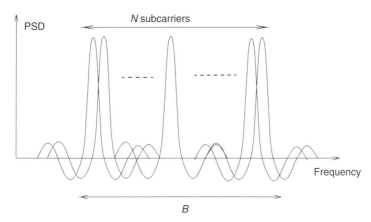

Figure 4.3 OFDM symbol presentation in the frequency domain

using several carriers, called also subcarriers, to modulate these substreams. The basis of a MCM modulation is illustrated in Fig. 4.5. The first systems using MCM were military HF radio links in the 1960s. Orthogonal Frequency Division Multiplexing is a special form of MCM with densely spaced subcarriers and overlapping spectra, as shown by the OFDM symbol representation in the frequency domain in Fig. 4.3. To allow an error-free reception of OFDM signals, the subcarriers' waveforms are chosen to be orthogonal to each other. Compared to modulation methods such as Binary Phase Shift Keying (BPSK) or Quadrature Phase Shift Keying (QPSK), OFDM transmits symbols that have relatively long time duration, but a narrow bandwidth. In the case of a symbol duration which is less than or equal to the maximum delay spread, as is the case with the other modulations, the received signal consists of overlapping versions of these transmitted symbols or Inter-Symbol Interference (ISI). Usually, OFDM systems are designed so that each subcarrier is narrow enough to experience frequency-flat fading. This also allows the subcarriers to remain orthogonal when the signal is transmitted over a frequency-selective but time-invariant channel. If an OFDM modulated signal is transmitted over such a channel, each subcarrier undergoes a different attenuation. By coding the data substreams, errors which are most likely to occur on severely attenuated subcarriers are detected and normally corrected in the receiver by the mean of forward error correcting codes.

In spite of its robustness against frequency selectivity, which is seen as an advantage of OFDM, any time-varying character of the channel is known to pose limits to the system performance. Time variations are known to deteriorate the orthogonality of the subcarriers; [Cimi85]. In this case, the Inter-Carrier Interference (ICI) appears because the signal components of a subcarrier interfere with those of the neighboring subcarriers.

By transmitting information on N subcarriers, the symbol duration of an OFDM signal is N times longer than the symbol duration of an equivalent single-carrier signal. Accordingly, ISI effects introduced by linear time dispersive channels are minimized. However, to eliminate the ISI completely, a guard time is inserted with a duration longer than the duration of the impulse response of the channel. Moreover, to eliminate ICI, the guard time is cyclically extended. It is to be noted that, in the presence of linear time dispersive channels, an appropriate guard time avoids ISI but not ICI, unless it is cyclically extended [Rodr02]. For this reason a guard time with T_{cp} duration is added to the OFDM

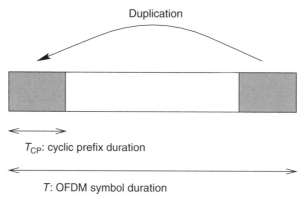

Figure 4.4 Adding the cyclic prefix by duplicating the first part of the original symbol

symbol, and in order to build a kind of periodicity around this OFDM symbol the content of this guard time is duplicated from the first part of the symbol, as represented in Fig. 4.4. In this case, the guard time becomes the cyclic prefix (CP).

The insertion of the appropriate cyclically extended guard time eliminates ISI and ICI in a linear dispersive channel; however, this introduces also a loss in the signal-to-noise ratio (SNR) and an increase of needed bandwidth; [Rodr02]. The SNR loss is given by Eq. (4.1).

$$\text{SNR}_{\text{loss}}(\text{dB}) = 10 \log \frac{T}{T - T_{\text{CP}}} \quad (4.1)$$

and the bandwidth expansion factor is given by

$$\varepsilon_B = \frac{T}{T - T_{\text{CP}}} \quad (4.2)$$

4.2.1.2 Generation of OFDM Signals

The generation of the OFDM symbols is based on two principles. First, the data stream is subdivided into a given number of substreams, where each one has to be modulated over a separate carrier signal, called subcarrier. The resulting modulated signals have to be then multiplexed before their transmission. Second, by allowing the modulating subcarriers to be separated by the inverse of the signaling symbol duration, independent separation of the frequency multiplexed subcarriers is possible. This ensures that the spectra of individual subcarriers are zeros at other subcarrier frequencies, as illustrated in Fig. 4.3, consisting of the fundamental concept of the orthogonality and the OFDM realization. Figure 4.5 shows the basic OFDM system [Cimi85]. The data stream is subdivided into N parallel data elements and are spaced by $\Delta t = 1/f_s$, where f_s is the desired symbol rate. N serial elements modulate N subcarrier frequencies which are then frequency division multiplexed. The symbol interval has now been increased to $N \Delta t$ which provides robustness to the delay spread caused by the channel. Each one of two adjacent subcarrier frequencies are then spaced by the interval formulated by Eq. (4.3).

$$\Delta f = \frac{1}{N \cdot \Delta t} \quad (4.3)$$

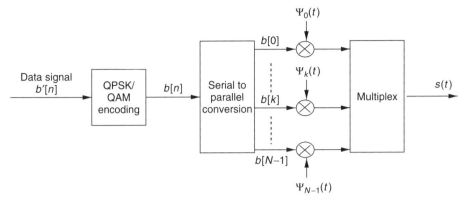

Figure 4.5 Basic OFDM transmitter

This ensures that the subcarrier frequencies are separated by multiples of $1/T$ so that the subcarriers are orthogonal over a symbol duration in the absence of distortions. It is to be noted that T in this phase is the OFDM symbol duration to which the cyclic period T_{cp} is not yet added.

According to the basic OFDM realization, the transmitted signal $s(t)$ can be expressed by

$$s(t) = \sum_{k=0}^{N-1} \sum_{l=-\infty}^{\infty} b_l[k] \psi_k(t - lT) \tag{4.4}$$

with the pulse having the function $p(t)$ and $f_k = k/T$, each subcarrier can be formulated by

$$\psi_k(t) = p(t) \cdot e^{j2\pi f_k t} \tag{4.5}$$

The basis $\{\psi_0, \psi_1, \psi_{N-1}\}$ is orthogonal, therefore

$$\int_0^T \psi_k(t) \psi_i^*(t)\, dt = \begin{cases} 1, & \text{if } i = k \\ 0, & \text{if } i \neq k \end{cases} \tag{4.6}$$

Then the transmitted signal can be expressed as

$$s(t) = \sum_{k=0}^{N-1} \sum_{l=-\infty}^{\infty} b_l[k] p(t - lT) \cdot e^{j2\pi f_k t} \tag{4.7}$$

By sampling at a rate $T_S = T/N$

$$x[n] = \sum_{k=0}^{N-1} \sum_{l=-\infty}^{\infty} b_l[k] \prod_N [nT_S - lNT_S] \cdot e^{j2\pi k n T_S/(NT_S)} \tag{4.8}$$

$$x[n] = \sum_{k=0}^{N-1} \sum_{l=-\infty}^{\infty} b_l[k] \prod_N [n - lN] \cdot e^{j2\pi k n/N} \tag{4.9}$$

with

$$\prod_N [n - lN] = \begin{cases} 1, & \text{for } (lN < n \leq (l+1)N) \\ 0, & \text{otherwise} \end{cases} \quad (4.10)$$

the signal can be presented in the form

$$x[n] = \sum_{l=-\infty}^{\infty} \prod_N [n - lN] \cdot \sum_{k=0}^{N-1} b_l[k] e^{j2\pi kn/N} \quad (4.11)$$

$$x[n] = \sum_{l=-\infty}^{\infty} \prod_N [n - lN] \cdot \text{IDFT}(b_l, n) \quad (4.12)$$

where IDFT is Inverse Discrete Fourier Transform.

From this presentation of an OFDM modulated signal, it can be deduced that for the generation of the OFDM signals $x[n]$ an IDFT block processing is required. The OFDM signal generation can be further optimized by calculating the IDFT of the original signals by the mean of the Inverse Fast Fourier Transform (IFFT). For the cyclic extension of the OFDM symbol, the last T_{cp} samples of the IFFT block output are inserted at the start of the OFDM symbol. At the receiver side, the first T_{cp} samples of the OFDM symbol have to be then discarded, as shown in Fig. 4.6.

4.2.1.3 Realization of OFDM System

The previous section has shown that the generation of the OFDM symbol can be realized through the IFFT/IFF processing block to which the mapped original data is applied. However, several complementary operations have to achieved and applied to the information bits before they are submitted to the IFFT processing, as illustrated by Fig. 4.6.

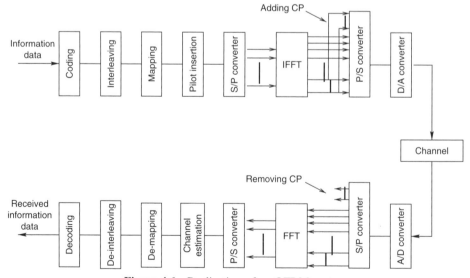

Figure 4.6 Realization of an OFDM system

The coding of the original information is a primordial step to make the transmission over the real channels possible, and this is because of the distortion. The interleaving of the encoded information should help avoid the long error bursts that limit the capability of the error correcting codes for detection and correction of errors. In more complex OFDM system realization, the so-called bit-loading procedure is applied. With this bit-loading, the amount of information (or bits) sent over a given subcarrier depends on the quality of this subcarrier. In this case, the bit rate realized over the subcarriers that are strongly affected by the disturbances is lower than the bit rate realized over the clean subcarriers.

The mean functionality required for the realization of an OFDM system can be summarized as follows:

Coding/Decoding and Interleaving/De-interleaving
At the transmitter side and before modulating the information signal, a channel coding is used so that the correctly received data of the relatively strong subcarriers corrects the erroneously received data of the relatively weak subcarriers. A set of channel coding schemes have been investigated for application within OFDM systems including block codes [NeePr00], convolutional codes [RohlMa99] and turbo codes [Somm02, BahaSa99]. Furthermore, the occasional deep fades in the frequency response of the transmission channel cause some groups of subcarriers to be less reliable than other groups and hence cause bit errors to occur in bursts rather than independently. Since channel coding schemes are normally designed to deal with independent errors and not with error bursts, the interleaving technique is used to guarantee this independence by effecting randomly scattered errors. For this reason, in the transmitter and after the coding, the bits are randomly permuted in such a way that adjacent bits are separated by several number of bits. At the receiver side, before the decoding, the de-interleaving is performed in order to get the original ordering of the bits. The interleaving function can be realized by block or convolution interleavers [BahaSa99]. A detailed discussion of the forward error correction (FEC) and interleaving classes is given in Sec. 4.3.

Mapping/De-mapping
After coding and interleaving, the bits to be conveyed in the l-th OFDM time slot and over the k-th OFDM subcarrier are mapped to a convenient modulation symbol, $S_{l,k}$. This mapping can be carried out with or without differential encoding. With no differential encoding, the data bits are directly mapped to the complex modulation symbols. Generally, this encoding is realized either by M-ary Phase Shift Keying (M-PSK) or by M-ary Quadrature Amplitude Modulation (M-QAM). In Fig. 4.7, a Gray encoded 16-PSK and 16-QAM signal constellation is illustrated, where binary words are assigned to adjacent symbol states and differ by only one digit.

With differential encoding, the data bits are not directly mapped to the complex modulation symbols $S_{l,k}$, but rather to the quotient $B_{l,k}$ of two successive complex modulation symbols, either in time direction or in frequency direction [Rodr02]. If the encoding is in the time direction, then

$$S_{l,k} = S_{l-1,k} \times B_{l,k} \tag{4.13}$$

and to initialize this differential mapping process each subcarrier of the first OFDM symbol conveys a known reference value. If encoding is performed in the frequency direction, then

$$S_{l,k} = S_{l,k-1} \times S_{l,k} \tag{4.14}$$

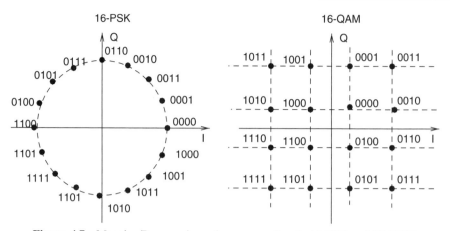

Figure 4.7 Mapping/De-mapping scheme according to 16-PSK and 16-QAM

and for the initialization of this differential encoding the first subcarrier of each OFDM symbol conveys a known reference value.

At the receiver and before the de-interleaving and decoding, the received modulation symbol $R_{l,k}$ is de-mapped to yield the bits conveyed in the l-th OFDM time slot and the k-th OFDM subchannel. Coherent detection or differential detection can be employed, according to the mapping scheme used at the transmitter, no differential or differential encoding, respectively. For mapping without differential encoding, the coherent detection is used, whereby the decision is based on the quotient $D_{l,k}$, [Rodr02], given by

$$D_{l,k} = \frac{R_{l,k}}{\hat{H}_{l,k}} \approx S_{l,k} + \frac{N_{l,k}}{\hat{H}_{l,k}} \qquad (4.15)$$

where $\hat{H}_{l,k}$ is an estimate of the channel transfer factor $H_{l,k}$ and $N_{l,k}$ is the component of the white additive Gaussian noise superposed to the transmitted symbol. Such an estimation is necessary to identify the amplitude and phase reverences of the constellation in each OFDM subcarrier so that the complex data symbols can be correctly demodulated. This simple equalization operation consist of the principal advantage of the OFDM receivers. Essentially by transmitting the original data over multiple narrowband subcarriers, the overall frequency-selective channel is transformed into a set of flat fading channels whose effect is only to introduce a random attenuation/phase shift in each OFDM subcarrier. Therefore, an OFDM channel equalizer corresponds to a bank of complex multipliers.

In the case of differential encoding, the differential detection must be used at the reception to get back the modulated symbols. If the differential coding was achieved in the time direction, then the differential detection is realized by comparing the information on the same subcarrier in consecutive OFDM symbols and the decision is based on the quotient [Rodr02]:

$$D_{l,k} = \frac{R_{l,k}}{R_{l-1,k}} = \frac{S_{l-1,k} B_{l,k} H_{l,k} + N_{l,k}}{S_{l-1,k} H_{l-1,k} + N_{l-1,k}} \qquad (4.16)$$

If the differential encoding was performed in the frequency direction, then the differential detection is performed by comparing the information on consecutive subcarriers in the same OFDM symbol and the decision is based on the following quotient

$$D_{l,k} = \frac{R_{l,k}}{R_{l,k-1}} = \frac{S_{l,k-1} B_{l,k} H_{l,k} + N_{l,k}}{S_{l,k-1} H_{l,k-1} + N_{l,k-1}} \quad (4.17)$$

By comparing the differential and the nondifferential detection methods, the differential schemes are very robust to residual phase offsets caused by a symbol timing offset or a non-perfect phase lock between the transmitter up-converter oscillator and the receiver down-converter oscillator. Moreover, differential schemes are realizable by simpler receiver implementations because no channel estimation is necessary, in contrast to the nondifferential schemes. However, in the presence of noise, the differential detection shows up to 3-dB degradation in the SNR when compared to the ideal coherent detection [Proa95].

Pilot Insertion/Channel Estimation
In the case of the coherent detection system, a channel estimate is necessary. This estimate is important to identify the amplitude and the phase reference of the mapping constellation in each subcarrier so that the complex data symbols can be de-mapped correctly. Channel estimation in OFDM systems requires the insertion of known symbols or pilot structure into the OFDM signal. These known symbols yield point estimates of the channel frequency response and an interpretation operation that yields the remaining points of the channel frequency response from the point estimates. The performance of the estimator depends strongly on how the pilot information is transmitted.

A typical two-dimensional pilot structure is investigated in [Rodr02]. This structure is adequate, since the channel can be viewed as a two-dimensional signal, in time and in frequency, sampled at the pilot positions, whereby also the two-dimensional sampling theorem imposes limits on the density of pilots to obtain an accurate representation of the channel. Essentially, the coherence time of the channel dictates the minimum separation of the pilots in the time direction and the coherence bandwidth of the channel dictates the minimum separation of the pilots in the frequency domain. In the pilot insertion, the higher the density of pilot symbols the better the accuracy. However, the higher the density of pilot symbols, the higher the loss in SNR and/or data rate [BahaSa99].

4.2.2 Spread-Spectrum Modulation

4.2.2.1 Principles of Spread Spectrum

Spread spectrum is a type of modulation that spreads data to be transmitted across the entire available frequency band, in excess of the minimum bandwidth required to send the information. The first spread-spectrum systems were designed for wireless digital communications, specifically in order to overcome the jamming situation, that is, when an adversary intends to disrupt the communication. To disrupt the communication, the adversary needs to do two things; first to detect that a transmission is taking place and sccond to transmit a jamming signal that is designed to confuse the receiver. Therefore, a spread-spectrum system must be able to make these tasks as difficult as possible. Firstly, the transmitted signal should be difficult to detect by the adversary, and for this reason the transmitted spread-spectrum signal is mostly called noise-like signal. Secondly, the signal should be difficult to disturb with a jamming signal.

Spread spectrum originates from military needs and finds most applications in hostile communications environments; such is the case in the PLC environments. Its typical applications are the cordless telephones, wireless LANs, PLC systems and cable replacement systems such as Bluetooth. In some cases, there is no central control over the radio resources, and the systems have to operate even in the presence of strong interferences from other communication systems and other electrical and electronic devices. In this case, the jamming is not intentional, but the electromagnetic interferences may be strong enough to disturb the communication of the nonspread spectrum systems operating in the same spectrum.

The principle of the spread spectrum is illustrated in Fig. 4.8, where the original information signal, having a bandwidth B and duration T_S, is converted through a pseudo-noise signal into a signal with a spectrum occupation W, with $W \gg B$. The multiplicative bandwidth expansion can be measured by a spread-spectrum parameter called *Spreading Factor* (SF). For military applications, the SF is between 100 to 1000, and in the UMTS/W-CDMA system the SF lies between 4 and 256. This parameter is also known as "spreading gain" or "processing gain" and is defined by Eq. (4.18).

$$G = \frac{W}{B} = W \cdot T_S \qquad (4.18)$$

Among the several advantages of spread-spectrum technologies, one can mention the inherent transmission security, resistance to interference from other systems, redundancy, resistance to multipath and fading effects. The common speed spread-spectrum techniques are Direct Sequence (DS), Frequency Hopping (FH), Time Hopping (TH), and the Multi-Carrier (MC). Of course, it is also possible to mix these spread-spectrum techniques to

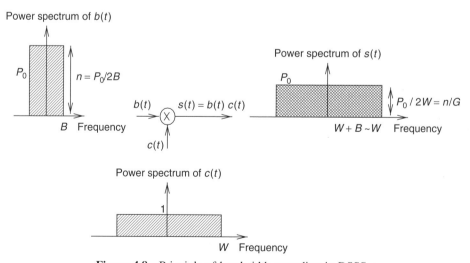

Figure 4.8 Principle of bandwidth spreading in DSSS

form hybrids that have the advantages of different techniques. We focus in this paragraph only on DS and HF. The DS is an averaging type system where the reduction of interference takes place because the interference can be averaged over a large time interval. The FH and TH systems are avoidance systems. Here, the reduction in interference occurs because the signal is made to avoid the interference for a large fraction of time.

4.2.2.2 Direct Sequence Spread Spectrum

Direct Sequence Spread Spectrum (DSSS) is the most applied form of the spread spectrum in several communications systems. To spread the spectrum of the transmitted information signal, the DSSS modulates the data signal by a high rate pseudorandom sequence of phase modulated pulses before mixing the signal up to the carrier frequency of the transmission system.

In the DSSS transmitter illustrated in Fig. 4.9, the information bit stream $b[n]$, which has a symbol rate $1/T_b$ and an amplitude from the set $\{-1, +1\}$, is converted into an electrical signal $b(t)$ through a simple Pulse Modulation Amplitude (PAM), generating a pulse train $\Pi_{T_b}(t)$. To spread the spectrum of the information signal $b(t)$, it is then multiplied by an unique high rate digital spreading code $c(t)$ that has many zero crossings per symbol interval with period T_c. For the generation of the spreading signal $c(t)$, first a code sequence $c[m]$ is generated by a Pseudo-Noise Sequence (PNS) generator with a frequency $1/T_c$ and then modulated through PAM with plus train $\Pi_{T_b}(t)$.

Different single-carrier modulations can be used to push the spread signal to the high frequency, such as BPSK and QPSK [Wong02], or the M-PSK [Meel99a]. By considering the DSSS transmitter based on BPSK modulation in Fig. 4.9, the signal carrier has a peak amplitude $(2E_b/T_b)^{1/2}$, where E_b is the energy per information bit. Then the transmitted signal $s(t)$ can be written as [StroOt02]

$$s(t) = \sqrt{\frac{2E_b}{T_b}} \cos(2\pi f_c t) b(t) c(t) \qquad (4.19)$$

where the data signal $b(t)$ is defined as

$$b(t) = \sum_{n=-\infty}^{\infty} b[n] \prod_{T_b}(t - nT_b) \qquad (4.20)$$

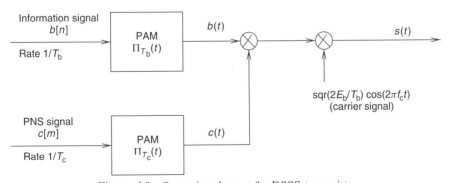

Figure 4.9 Synoptic scheme of a DSSS transmitter

and the wave form of the spreading code, which is a baseband signal, is defined by

$$c(t) = \sum_{m=-\infty}^{\infty} c[m] \prod_{T_c}(t - mT_c) \qquad (4.21)$$

where $\Pi_T(t)$ denotes an unit amplitude rectangular pulse with a duration of T.

By taking $1/T_c = N/T_b$, after the modulation the transmitted signal has a bandwidth of $2N/T_b$. This means that the bandwidth of the transmitted signal is N times wider than the bandwidth of the original information signal. Then, the spreading factor is equal to N.

At the receiver side, demodulation and a de-spreading operation are realized to recuperate the original signal. From a modulation perspective, the receiver is just a down-mixing stage followed by a filter which is matched to consecutive T_b-segments of $c(t)$, a so-called code matched filter. The multiplication by the demodulating signal with frequency f_c consists in pushing the signal back to its baseband form. Then a code sequence $c(t)$ identical to the one generated in transmitter have to be generated at the receiver and multiplied with the baseband signal. If a good synchronization between the two codes sequences is realized, their correlation, called also autocorrelation (see Sec. 5.2.3), will be equal to one. In this case, after submitting the baseband signal to a correlator, we get, at its output, a signal $\hat{b}(t)$, which normally is similar to $b(t)$. The obtained signal is then sampled at a rate $1/T_b$ and a decision or estimation about the original amplitude of sample, either $+1$ or -1, is made in order to build the original bit stream $b[n]$. The synoptic scheme of the receiver where a matched filter is implemented with a correlator is illustrated in Fig. 4.10. There are other possible solution schemes that can be used at the receiver side according to the techniques used at the transmitter side, such as receivers based on "chip matched filter" with an arbitrary chip waveform [StroOt02].

4.2.2.3 Frequency Hopping Spread Spectrum

In a Frequency Hopping Spread-Spectrum system (FHSS) the signal frequency is constant for specified time duration, referred to as a time chip T_c. The transmission frequencies are then changed periodically. Usually, the available band is divided into nonoverlapping frequency "bins". The data signal occupies one and only one bin for a duration T_c and hops to another bin afterward. It is frequently convenient to categorize frequency hopping system as either "fast-hop" or "slow-hop", since there is a considerable difference in the performance for these two types of systems. A fast-hop system is a system in which the

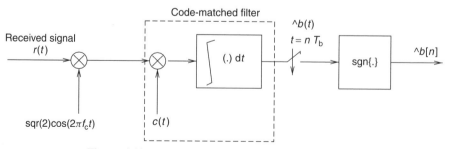

Figure 4.10 A DSSS receiver based on matched filter

frequency hopping takes place at a rate $1/T_h$, which is greater than the message bit rate $1/T_s$, as illustrated in Fig. 4.11 using a 4-ary FSK modulation and where T_h is taken equal to $T_s/2$. In a slow-hop system, the hop rate is less than the message bit rate, for example $1/T_h$ is equal to $1/2T_s$ as illustrated in Fig. 4.11 also.

The block diagram of a fast-hop FHSS transmitter and its corresponding receiver are presented in Fig. 4.12 and Fig. 4.13 respectively. In the FHSS system, the modulation schemes, such as M-ary FSK, which allow noncoherent detection, are usually employed for the data signals, because it is practically difficult to build coherent frequency synthesizers [Wong02]. According to the generated pseudorandom sequence code, the frequency synthesizer generates a signal with a frequency among a predefined set of possible frequencies, which has to carry the baseband signal over the transmission channel.

For M-ary FSK, the data signal can be expressed as

$$b(t) = \sqrt{2P} \cdot \sum_{n=-\infty}^{\infty} \prod_{T_b}(t - nT_b) \cos(2\pi f_n t + \phi_n) \qquad (4.22)$$

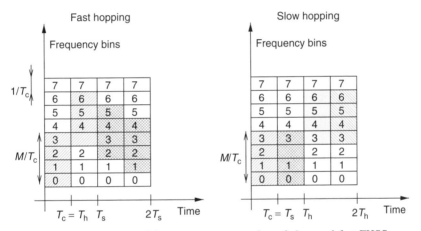

Figure 4.11 Time and frequency representation of slow and fast FHSS

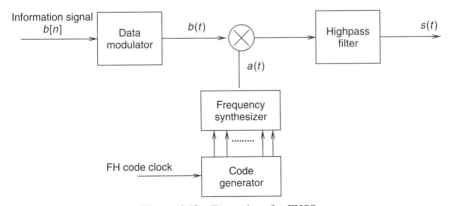

Figure 4.12 Transmitter for FHSS

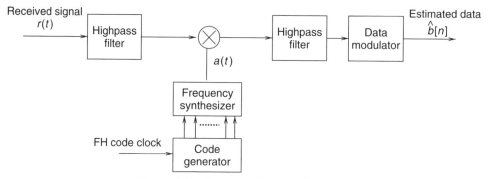

Figure 4.13 Receiver for a FHSS system

where $f_k \in \{f_{s0}, f_{s1}, \ldots, f_{sM-1}\}$ and P is the average transmitted power. The frequency synthesizer outputs a hopping signal

$$a(t) = 2 \cdot \sum_{m=-\infty}^{\infty} \prod_{T_c}(t - mT_c) \cos(2\pi f'_m t + \phi'_m) \tag{4.23}$$

where $f'_m \in \{f_{c0}, f_{c1}, \ldots, f_{cL-1}\}$. In this case, there are L frequency bins in that FHSS system.

Let $T_b = N * T_c$ be the constraint for fast hopping, which becomes $T_c = N * T_b$ in case of slow hopping. The transmitted signal in a fast hopping FHSS is given by [Wong02]

$$s(t) = \sqrt{2P} \sum_{m=-\infty}^{\infty} \prod_{T_c}(t - mT_c - \Delta) \cos[(2\pi f_{\lfloor m/N \rfloor} + 2\pi f'_m)(t - \Delta) + \phi_{\lfloor m/N \rfloor} + \phi'_m] \tag{4.24}$$

and in the case of slow hopping, $T_b \leq T_c$, the transmitted signal is

$$s(t) = \sqrt{2P} \sum_{n=-\infty}^{\infty} \prod_{T_b}(t - nT_b - \Delta) \cos[(2\pi f_n + 2\pi f'_{\lfloor n/N \rfloor})(t - \Delta) + \phi_n + \phi'_{\lfloor n/N \rfloor}] \tag{4.25}$$

where $\lfloor x \rfloor$ is the largest integer which is smaller than or equal to x and Δ is a uniform random variable on $[0, T_b)$. The requirement of orthogonality for the FSK signals forces the separation between the adjacent FSK symbol frequencies be at least $1/T_c$ for fast hopping, and $1/T_b$ for slow hopping. Therefore, the minimum separation between adjacent hopping frequencies is M/T_c for fast hopping, and M/T_b for slow hopping.

At the FHSS receiver side, the main task is to regenerate a pseudorandom sequence that must be similar to the one generated at the transmitter, and according to which the modulation of the signal in the high frequency was achieved. This should allow a correct demodulation of the transmitted signal. However, it is important to note that another demodulation has to follow, in our example, according to the M-ary FSK. Then the recuperated signal has one of the M possible frequencies and this should allow a correct estimation of the value of $b[n]$ at each time period T_b.

4.2.2.4 Comparison of DSSS and FHSS

The comparison can be achieved according to different evaluation parameters, such as the spectral density reduction, interference susceptibility, capacity, and so on. Furthermore, the choice of the suitable scheme according to the system needs is based on parameters that are linear or inversely dependent on each other. Both DSSS and FHSS reduce the average power spectral density of a signal. The way they do it is fundamentally different and has serious consequences for other users. For an optimal system realization, the objectives are to reduce both transmitted power and power spectral density, to keep them from interfering with other users in the band. DSSS spreads its energy by phase-chopping the signal so that it is continuous only for brief time intervals (or chip). Therefore, instead of having all the transmitted energy concentrated in the data bandwidth, it is spread out over the spreading bandwidth. The total power is the same, but the spectral density is lower. Of course, more channels are interfered with than before, but at a much lower level. Furthermore, if the spread signal comes in under the noise level of most other users, it will not be noticed. Traditional FHSS signals lower only their "average" power spectral density hopping over many channels. But during one hop, a FHSS signal appears to be a narrow band signal, with a higher power spectral density.

The interference susceptibility is another important parameter which allows the system to operate properly. In DSSS receivers, the de-spreading operation consists in multiplying the received signal by a local replica of the spreading code. This correlates with the desired signal to push it back to the data bandwidth, while spreading all other noncorrelating signals. After the de-spread signal is filtered to the data bandwidth, most of the noise is outside this new narrower bandwidth and is rejected. This helps only with all types of narrowband and uncorrelated interference, and it has no advantage for wideband interference since spread noise is still noise and the percentage that falls within the data bandwidth is unchanged.

The FHSS signal is agile and does not spend much time on any one frequency. When it hits a frequency that has too much interference, the desired signal is lost. In a packet switched system, this results in a retransmission, usually over a clearer channel. In a fast enough FHSS system, the portion of lost signal may be recovered by using a FEC. Other parameters and comparisons of the DSSS and FHSS are listed in Tab. 4.1, from which it becomes clear that the DSSS shows more advantages than the FHSS systems [Meel99a].

4.2.3 Choice of Modulation Scheme for PLC Systems

Several investigations have been carried out to find suitable OFDM implementations for PLC networks. In order to avoid hard degradation of OFDM signal over the transmission channel, which is caused by the frequency-selective fading, a method for subcarriers power control is proposed in [NomuSh01]. This solution consists of controlling the transmission power of each subcarrier of OFDM signal in order to maximize the average SNR of each subcarrier of the received signal. This controlling is so flexible that the total transmitted power is not increased. Further improvement of such controlling is possible by spreading the parallel substreams at the output of the serial-to-parallel converter output [NishNo02, NishSh03]. An OFDM system which subdivides the original information into three parallel data groups, where each group is mapped either according to BPSK or QPSK and coded according to Reed–Solomon code or convolutional code, is also

Table 4.1 Comparison of the advantages (+) and drawbacks (−) of DSSS and FHSS

	DSSS	FHSS
Spectral density and interference generation	+ Reduced with processing gain + Continuous spread of the transmitted signal power gives minimum interference	+ Reduced with processing gain − Only the average power of the transmitted signal is spread, and this gives less interference reduction
Transmission	+ Continuous, broadband	− Discontinuous, narrowband
Interference susceptibility	+ Narrowband interference in the same channel is reduced by the processing gain	− Narrowband interference in the same interference is not reduced + Hopping makes transmission on usable channels possible
Higher data rates	+ The data rate can be increased by increasing the clock rate and/or the modulation complexity (multilevel)	− A wider bandwidth is needed but is not available (it would cut the number of channels to hop in)
Real time (voice)	+ No timing constraints − If a station is jammed, it is jammed until the jammer goes away	− If a channel is jammed, the next available transmission time on a clear channel may be T_c duration away
Synchronization	+ Self-synchronization	− Many channels to search
Implementation	− Complex baseband processing	+ Simple analog limiter/discriminator receiver

investigated in [KuriHa03]. Performances of OFDM system were also investigated under different noise scenarios, especially under the impulsive noise, which is considered the dominating noise in PLC environment, [ShirNo02, MatsUm03].

Spread-spectrum modulation techniques, with direct sequence or frequency hopping, were investigated to be implemented in PLC physical layer. For example, in [FerrCa03], a so-called "low complexity all-digital DSSS transceiver" is proposed, which is based on a delay-locked loop for clock recovery and on a phase recovery that is implicit in the timing synchronization. An "iterative detection algorithm" for M-ary spread-spectrum system over a noisy channel is investigated and this shows a remarkable improvement of the detection performance for M-ary systems [UmehKa02]. However, the main drawback of the spread-spectrum technique is the relative lower realizable bit rate, in comparison with OFDM systems. This makes any decision about the modulation to be adopted for a PLC system more difficult. By deciding for a given modulation, the system designer must know which performances have the higher priority for him and which ones have less importance. Besides the high realizable bit rates, the OFDM systems also show a high robustness against the channel distortions, a flexibility in avoiding the strongly affected channels and an optimal bandwidth utilization by the usage of the slightly disturbed channels through the bit-loading procedure. The main advantage of the spread spectrum is its electromagnetic compatibility, by the radiation of weak electromagnetic fields in the environment [Dost01a].

4.3 Error Handling

4.3.1 Overview

PLC networks operate with a signal power that has to be below a limit defined by the regulatory bodies (Sec. 3.3). On the other hand, the signal level has to keep data transmission over PLC medium possible. That means, there should be a certain SNR (*Signal- to-Noise Ratio*) level in the network making communications possible. As long as the SNR is sufficient to avoid the disturbances in the network, the error handling mechanisms do not have to act; for example, if the SNR is sufficient to avoid an influence of the background noise in a PLC network.

More difficulties in PLC transmission systems are caused by impulsive noise, which has much higher power than the background noise. In this case, the SNR is not enough to overcome the disturbances and the resulting transmission errors. However, if the duration of a disturbance is short enough, the physical layer can deal with it, as described in Sec. 4.2. On the other hand, if the noise impulses are longer, additional mechanisms for error handling have to be applied: mechanisms for error correction and retransmission mechanisms for short-term disturbances and capacity reallocation mechanisms for long-term disturbances.

In many transmission systems, forward error correction and interleaving mechanisms are applied to cope with the disturbances [DaviBe96]. In this case, the transmission systems are able to manage a situation when a number of bits are damaged and, in spite of that, to correct the data contents. The usage of the FEC mechanism gives rise to an overhead, which takes a portion of the network transmission capacity; for example, about 50% overhead is used for the FEC in the GSM system, which improves BER (*Bit Error Rate*) values from 10^{-3} (pure wireless transmission channel) to 10^{-6} [Walke99]. Particular methods for FEC to be applied in PLC networks are the point of current and future research works [Zimm00]. We present an overview of currently considered FEC and interleaving mechanisms for PLC in Sec 4.3.2 and 4.3.5 respectively.

In spite of the applied FEC mechanisms and the ability of communications systems to avoid different kinds of disturbances, it is still possible that the transmitted data may be damaged. In the case of errors, the damaged data has to be retransmitted by an ARQ (Automatic Repeat reQuest) mechanism. The application of ARQ can reduce error probability to a very low value and it is only limited by the remaining error probability of CRC (*Cyclic Redundancy Check*) code used for error recognition, or error tolerance specified by a particular application. To deal with disturbances, various communications systems apply a so-called hybrid ARQ/FEC solution, a combination of ARQ and FEC mechanisms (e.g. [KousEl99, Joe00]), which is also expected to be used in PLC networks. The application of ARQ is suitable for data transmission without delay requirements. However, for time-critical services, such as telephony, ARQ adds additional delays that may be not acceptable. The basic variants of ARQ mechanisms are described in Sec. 4.3.4.

The ARQ mechanisms deal with a relatively short duration of the disturbances (some milliseconds) that occur on one or several data units. On the other hand, so-called long-term disturbances (e.g. caused by narrowband noise produced by short-wave radio stations) make one or more transmission channels unavailable for a longer time. In this case, the ARQ mechanism would constantly repeat the data, making the transmission inefficient. Because of that, long-time disturbed transmission channels should not be used for any transmission until the disturbance disappears. If a disturbed channel is currently used for the transmission, channel reallocation has to be carried out to allow the continuation of affected connections using other error-free channels. The possibility for implementation of reallocation mechanisms has to be also included in the features of the PLC MAC layer and they are considered in Sec. 5.4.3.

4.3.2 Forward Error Correction

Forward Error Correction (FEC) is a widely used method to improve the connection quality in digital communications and storage systems. The word "forward" in conjunction with error correction means the correction of transmission errors at the receiver side without needing any additional information from the transmitter. The main concept of FEC is to add a certain amount of redundancy to the information to be transmitted, which can be exploited by the receiver to correct transmission errors due to channel distortion and noise. Therefore, in the literature, the FEC coding is mostly described as channel coding. Shannon presented in his mathematical theory of communication that every transmission channel has a theoretical maximum capacity, which depends on the bandwidth and the signal-to-noise-ratio (SNR), as formulated by Eq. (4.26) [Shan49]. The capacity of implemented systems is mostly much smaller than the maximum possible value calculated by the theory. For this reason, the use of suitable codes has to allow further improvement in bandwidth efficiency.

Shannon's capacity theorem states that, for an AWGN channel, the maximum reliable, that is, error-free, transmission rate is given by

$$R \leq B \cdot \log_2 \left(1 + \frac{P}{N_0 B}\right) \quad (4.26)$$

where B represents the channel bandwidth, N_0 is the power spectral density of the noise, P the transmitted power and R is the communication bit rate in bits per second (bps). This expression can be rearranged to give the minimum E_b/N_0 required for reliable communication as a function of R/B:

$$\frac{E_b}{N_0} \geq \frac{2^{\frac{R}{B}} - 1}{\frac{R}{B}} \quad (4.27)$$

The minimum value for E_b/N_0 is obtained when R/B, called "bandwidth efficiency", approaches zero. This provides a lower bound for E_b/N_0 below which "reliable" communication is not possible. This is the "Shannon limit":

$$\frac{E_b}{N_0} > 10 \log_{10}(\log_e 2) \quad (4.28)$$

$$\frac{E_b}{N_0} = -1.6 \, \text{dB} \quad (4.29)$$

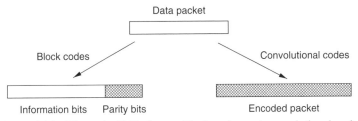

Figure 4.14 The main FEC classes: block codes and convolutional codes

For example, for a bandwidth efficiency of $R/B = 1$ bps/Hz, the limit for reliable communication is 0 dB.

The error correcting codes can be divided in two main families: block codes and convolutional codes, also called trellis codes, as illustrated in Fig. 4.14. Block codes add a constant number of parity bits to a block of information bits whose the length is constant, whereas convolutional codes generate a modified output bit stream with a higher rate than the input stream. In this section, we present these code families, including their principles, their properties and examples of realization. The turbo codes consist of a special subclass of the convolution codes that show high performances, and they will be discussed separately.

The various codes have different properties with respect to error correction performance and decoding complexity. Additionally, for a real system, design factors like block size and scalability have other practical constraints. However, channel codes should meet the following requirements, and/or try at least to realize a certain trade-off between them:

- Channel codes should have a high rate to maximize data throughput,
- Channel codes should have a good bit error rate performance at the desired SNR to minimize the energy needed for transmission,
- Channel codes should have low encode/decoder complexity to limit the size and cost of the transceiver, and
- Channel codes should introduce only minimal delays, especially in voice transmission, so that no degradation in signal quality is detectable.

4.3.2.1 Block Codes

When using block codes, the data to be transmitted is segmented into blocks of a fixed length k. To each block of the information message m, a certain amount of parity bits are added. The information bits and the parity bits together form the code words c of length n, as illustrated by Fig. 4.15, which shows a communications system coding the original information before submitting them to the modulation. The rate of a (n, k) block code is defined as $r = k/n$. Block codes might be separated into two main families: binary and nonbinary codes. Examples for binary codes are Cyclic, Hamming, Fire, Golay and BCH (Bose, Chaudhuri and Hocquenghem) codes [LinCo83]. The nonbinary codes work on symbols consisting of more than one bit. The most popular example is the Reed–Solomon (RS) codes, which are derived from binary BCH codes.

An (n, k) binary code, C, consists of a set of 2^k binary codes, each of length n bits, and a mapping function between message words and code words, as illustrated by the following example:

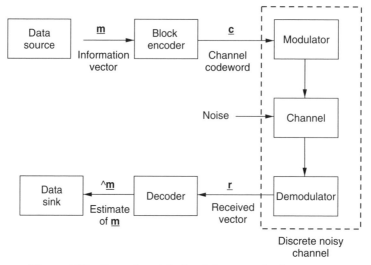

Figure 4.15 General model of coded communications system

Table 4.2 Example of mapping function of a binary (2,5) linear code

m	c
00	01100
01	10101
10	10111
11	11000

Binary (5,2) code with rate $r = 2/5$, where $C = \{01100, 10101, 10111, 11000\}$ and the mapping function defined by Tab. 4.2.

All the block codes used in the practice are linear. This means that the modulo-2 addition of two code words is also a valid code word [Schu99]. Linear block codes have several properties that are important for practical implementation. The codes can be defined in the form of a generator matrix and a parity check matrix. The syndrome concept can be used to detect and correct errors on the receiver side, as discussed below.

An (n, k) linear block code is defined by a generator matrix $\underline{G}$, such that the code word $\underline{c}$ for message $\underline{m}$ is obtained from Eq. (4.30) or Eq. (4.31), where modulo-2 arithmetic is used.

$$\underline{c} = \underline{m} \cdot \underline{G} \tag{4.30}$$

$$[c_1 \quad c_2 \quad \ldots c_n] = [m_1 \quad m_2 \quad \ldots m_k] \begin{bmatrix} g_{1,1} & g_{1,2} & \cdots & g_{1,n} \\ g_{2,1} & g_{2,2} & \cdots & g_{2,n} \\ \cdots & \cdots & \cdots & \cdots \\ g_{k,1} & g_{k,2} & \cdots & g_{k,n} \end{bmatrix} \tag{4.31}$$

The simple example of the linear block code is the Single Parity Check code, which is a $(k + 1, k)$ code defined by Eq. (4.32) and whose generator matrix G is formulated by Eq. (4.33).

$$c_k = m_1 \oplus m_2 \oplus \ldots \oplus m_k \tag{4.32}$$

$$\underline{G} = \begin{bmatrix} 1 & 0 & \ldots 0 & 1 \\ 0 & 1 & \ldots 0 & 1 \\ \ldots & \ldots & \ldots \ldots & . \\ 0 & 0 & \ldots 1 & 1 \end{bmatrix} \tag{4.33}$$

Furthermore, associated with very linear (n, k) code is a two-dimensional matrix called "parity check matrix", denoted by H with dimensions $(n - k)$ and n. This matrix is defined such that

$$\underline{GH}^T = \underline{0} \tag{4.34}$$

This matrix allows us to define the "syndrome" $\underline{s}$ of a received word $\underline{r}$ according to Eq. (4.35). The syndrome is of length $n - k$ bits.

$$\underline{s} = r\underline{H}^T \tag{4.35}$$

Then, if the received word does pertain to the code C, its syndrome is equal to zero as shown by Eq. (4.36), and therefore, no error is detected. In the other case where the syndrome is nonzero, the decoder has to take action to correct the errors. However, the capability of codes to correct the errors is limited, as described by Eq. (4.38), and in this case, the receiver has to request the retransmission of the code word through the ARQ mechanisms.

$$\underline{s} = r\underline{H}^T = c\underline{H}^T = m\underline{GH}^T = \underline{m0} = \underline{0} \tag{4.36}$$

The Hamming weight of a word is the number of 1's in the word, for example, $w_H(110110) = 4$. The Hamming distance between two words $\underline{a}$ and $\underline{b}$ is the number of positions in which $\underline{a}$ and $\underline{b}$ differ and is pointed out by $d_H(\underline{a}, \underline{b})$, for example, $d_H(01011, 11110) = 3$. The minimum distance of a code, C, is the minimum Hamming distance between any two different code words in C. The minimum Hamming distance can also be defined by Eq. (4.37). For example, for the code $C = \{00000, 01011, 10101, 11110\}$, the minimal hamming distance is $d_{\min} = 3$.

$$d_{\min} = \min\{d_H(\underline{a}, \underline{b}) | \underline{a}, \underline{b} \in C, \underline{a} \neq \underline{b}\} \tag{4.37}$$

The parameter $d_{\min}$ can be used to predict the error protection capability of a code. A block code with minimum distance $d_{\min}$ guarantees correcting all patterns of t or fewer errors, where t is upper bounded by $(d_{\min} - 1)/2$; [Lee00]. In this case, t is called "random-error correcting capability" of the code.

$$t = \lfloor (d_{\min} - 1)/2 \rfloor \tag{4.38}$$

or

$$t \leq (d_{\min} - 1)/2 \tag{4.39}$$

The main classes of the block codes that are widely used in the practice are the Hamming codes and the cyclic codes.

Hamming Codes

Hamming codes are a subclass of linear block codes that are able to correct exactly one error. For any positive integer $m \leq 3$, there exists a Hamming code with the following parameters:

- Code length: $n = 2^m - 1$
- Number of information symbols: $k = 2^m - m - 1$
- Number of parity symbols: $n - k = m$
- Error correction capability: $t = 1$, because $d_{min} = 3$.

Cyclic Codes

The cyclic codes are an important subclass of linear block codes, because the encoding and syndrome calculation can be realized by employing linear feed back shift registers. Cyclic codes are linear block codes with the additional constraint that every cyclic shift of a code word is also a code word, so that, if

$$\underline{c} = (c_0, c_1, c_2, \ldots, c_{n-1}) \in C$$

then

$$\underline{c}^{(1)} = (c_{n-1}, c_0, c_1, c_2, \ldots, c_{n-2}) \in C$$

where $\underline{c}^{(1)}$ is the right cyclic shift of $\underline{c}$.

Codes with this structure allow a simple implementation of the encoder and the syndrome calculator using shift registers, as illustrated in Fig. 4.16 and Fig. 4.17 respectively. Therefore, there is no need anymore for the complex matrix multiplications, and the cyclic codes are generally discussed in terms of polynomials. Every code word can be represented by a polynomial, as in Eq. (4.40).

$$\underline{c} = (c_0, c_1, c_2, \ldots, c_{n-1}) \Leftrightarrow c(X) = c_0 + c_1 X + c_2 X^2 + \cdots + c_{n-1} X^{n-1} \tag{4.40}$$

where $c_i = \{0, 1\}$ for binary cyclic codes.

The cyclic codes are defined by a polynomial generator of degree $n - k$, whose coefficient is $g_i = \{0, 1\}$ for binary cyclic codes, and is expressed as follows:

$$g(X) = 1 + g_1 X + g_2 X^2 + \cdots + g_{n-k-1} X^{n-k-1} + X^{n-k} \tag{4.41}$$

Then each message polynomial $m(X)$ is encoded to code polynomial $c(X)$, with

$$c(X) = m(X) g(X) \tag{4.42}$$

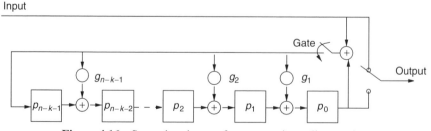

Figure 4.16 Synoptic scheme of a systematic cyclic encoder

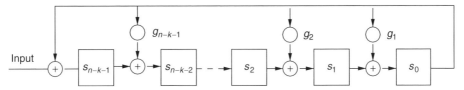

Figure 4.17 Syndrome calculation at the cyclic decoder, with $\underline{s} = (s_0, s_1, \ldots, s_{n-k-1})$

A general structure of a cyclic encoder based on the shift register, whose feedback coefficients are to be determined directly by the generating polynomial, is presented in Fig. 4.16. The generation of the code words is realized in four steps:

- *Step 1*: the gate is closed and the switch is set to position 1,
- *Step 2*: the k message bits are shifted in,
- *Step 3*: the gate is opened and the switch is set to position 2, and
- *Step 4*: the contents of the shift register are shifted out.

The syndrome calculation of systematic codes is also easily realized by the shift registers, according to the general scheme presented in Fig. 4.17. The operation of this syndrome calculator is also easy: we shift only the n received message bits, and the syndrome will be stored as contents of the shift registers, with $\underline{s} = (s_0, s_1, \ldots, s_{n-k-1})$.

As examples of the cyclic codes, one can mention the following widely used ones:

- Cyclic Redundancy Check (CRC) Codes:

These codes are often used for error detection with ARQ schemes. The most commonly used generator is that formulated by equation Eq. (4.43).

$$g(X) = 1 + X^2 + X^{15} + X^{16} \qquad (4.43)$$

- Bose–Chaudhuri–Hocquenghem (BCH) Codes:

    This is a large class of cyclic codes, where for any $m >= 3$ and $t >= 1$ there is a BCH code with
    Code length: $n = 2^m - 1$
    Number of parity symbols: $n - k =< mt$
    Minimum hamming distance: $d_{\min} = 2t + 1$
- Reed–Solomon (RS) Codes:

    The Reed–Solomon codes are nonbinary BHC codes, which work with symbols of k bits each [Schu99]. Message words consist of Km-bit symbols, and code words consist of Nm-bit symbols, where

$$N = 2^m - 1$$

The code rate is $R = K/N$

Reed-Solomon can correct up to t symbol errors, which makes it more adequate for correcting the error bursts, with

$$t = \left\lfloor \frac{1}{2}(N - K) \right\rfloor \qquad (4.44)$$

4.3.2.2 Convolution Codes

In the convolutional codes (also called trellis codes), the redundancy that must be added to allow error correction at the receiver is continuously distributed in the channel bit stream. Therefore, as opposed to the block codes, which operate on finite-length blocks of message bits, a convolutional encoder operates on continuous sequences of message symbols.

Let $\underline{a}$ denotes the message sequence with

$$\underline{a} = a_1 a_2 a_3 \ldots \tag{4.45}$$

and $\underline{c}$ denotes the code sequence of the form

$$\underline{c} = c_1 c_2 c_3 \ldots \tag{4.46}$$

At each clock cycle, a (n, k, m) convolutional encoder takes one message symbol of k message bits and produces one code symbol of n code bits. Typically, k and n are small integers (less than 5), with $k < n$. The parameter m refers to the memory requirement of the encoder. Increasing m improves the performance of the code, but this will also increase the decoder complexity. Therefore, the parameter m is typically less or equal to eight.

The basis for generating the convolutional codes is the convolution of the message sequences with a set of generator sequences. Let $\underline{g}$ denote a generator sequence of length $L + 1$ bits that can be presented by

$$\underline{g} = g_1 g_2 g_3 \ldots g_L \tag{4.47}$$

Let the convolution of $\underline{a}$ and $\underline{g}$ be $\underline{b} = b_1 b_2 b_3 \ldots$, with each output bit given by Eq. 4.48.

$$b_i = \sum_{l=1}^{L} a_{i-l} \cdot g_l \tag{4.48}$$

Different subclasses of the convolution codes can be realized according to the values assigned to their three parameters, namely n, k and m. We give below the general realization and/or examples of practical realization for the three main classes: $(2,1,m)$, $(n,1,m)$ and (n,k,m).

$(2, 1, m)$ Convolutional Codes

For a rate of 1/2 convolutional codes, two generator sequences, denoted by $\underline{g}^{(1)}$ and $\underline{g}^{(2)}$, are used. The two convolution output sequences are then $\underline{c}_i^{(1)}$ and $\underline{c}_i^{(2)}$, with

$$c_i^{(1)} = \sum_{l=1}^{L} a_{i-l} \cdot g_l^{(1)} \tag{4.49}$$

$$c_i^{(2)} = \sum_{l=1}^{L} a_{i-l} \cdot g_l^{(2)} \tag{4.50}$$

These two sequences are then multiplexed together to build up the code sequence given by Eq. (4.51).

$$\underline{c} = c_1^{(1)} c_1^{(2)} c_2^{(1)} c_2^{(2)} c_3^{(1)} c_3^{(2)} \ldots \tag{4.51}$$

The code is generated by passing the message sequence through an L-bit shift register, as illustrated in Fig. 4.18. This coder is of rate 1/2 because for each encoder clock cycle one message bit ($k = 1$) enters the encoder and simultaneously two code bits ($n = 2$) are produced. The memory m is in this case equal to L. Realization of a (2,1,2) convolution encoder with $\underline{g}^{(1)} = 101$ and $\underline{g}^{(2)} = 111$ is given in Fig. 4.19.

(n,1, m) Convolution Codes

Convolutional codes with rate $1/n$ can be designed by using n different generators. As an example of such encoders, Fig. 4.20 shows the synoptic scheme of a (3,1,3) encoder, with $\underline{g}^{(1)} = 1101$, $\underline{g}^{(2)} = 1110$ and $\underline{g}^{(3)} = 1011$. This code is of rate 1/3 because at each clock cycle one message bit ($k = 1$) enters the coder and three code bits ($n = 3$) are produced at the output. The memory m in this case is three bits.

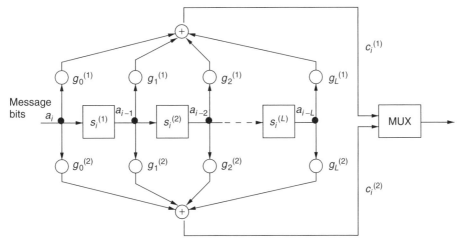

Figure 4.18 General model of a (2,1,m) convolutional encoder

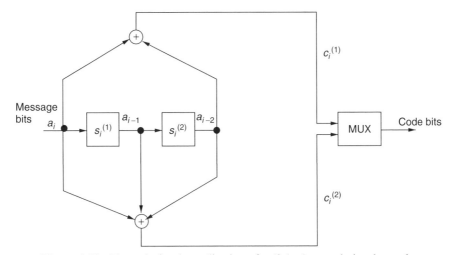

Figure 4.19 Example for the realization of a (2,1,m) convolutional encoder

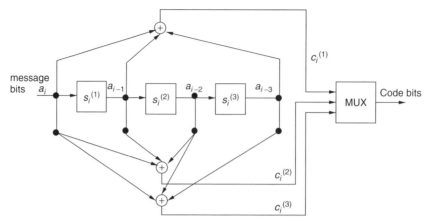

Figure 4.20 Example of a $(n, 1, m)$ convolutional encoder realization

(n,k,m) Convolution Codes

A k/n convolution encoder can be constructed by using multiple shift registers. The input sequence is demultiplexed into k separated streams, which are then passed through all the k shift registers. Therefore, one message, which is symbol k message bits, enters the encoder with each encoder clock cycle, and code symbol of n code bits is produced. As an example, Fig. 4.21 shows the structure of a (3,2,3) convolutional coder, whose generators are

$$\underline{g}^{(1,1)} = 100; \quad \underline{g}^{(2,1)} = 01$$
$$\underline{g}^{(1,2)} = 111; \quad \underline{g}^{(2,2)} = 11$$
$$\underline{g}^{(1,3)} = 001; \quad \underline{g}^{(2,3)} = 10$$

This code is pointed out as rate 2/3 because at each clock cycle two message bits ($k = 2$) enter the encoder and three code bits ($n = 3$) are then produced. The total required memory m is of three bits.

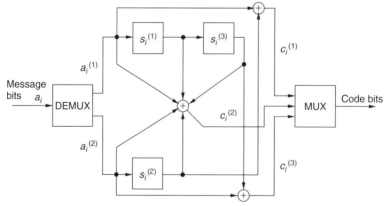

Figure 4.21 Example of realization of an (n,k,m) convolutional encoder

The decoding of convolutional codes is much more difficult than the encoding. The goal is to reconstruct the original bits sequence from the channel bit stream. According to [Schu99], there are three major methods to do this task: maximum likelihood decoding, sequential decoding and threshold decoding. The first of the three methods is commonly performed by the Viterbi algorithm, and was investigated to be implemented in PLC system [NakaUm03]. As an example for sequential decoding, the Fano algorithm is proposed. The three approaches differ in decoding complexity, delay and performance and the design will be a trade-off between these parameters. Furthermore, the Viterbi algorithm is an optimal maximum-likelihood sequence estimation algorithm for decoding convolutional codes by finding the most likely message sequence (message word) to have been transmitted based on the received word. In this case, the Viterbi minimizes the probability of a message word error. In [Mars03], the so-called Maximum a posteriori (MAP) decoding is discussed as alternative approach to decoding convolutional codes that is based on minimizing the probability of a message bit error.

4.3.2.3 Turbo Codes

Turbo coding was introduced first in 1993 by Berrou [BerrGl93]. Extremely impressive results were reported for a code with a long frame length that is approaching the Shannon channel capacity limit. Since its recent invention, turbo coding has evolved at an unprecedented rate and has reached a state of maturity within just a few years because of the intensive research efforts of the turbo coding community. As a result, turbo coding has also found its way into standard systems, such as the standardized third-generation (3G) mobile radio systems [SteeHa99] and is being discussed for adoption for the video broadcast systems standards. The turbo encoders are based on a given type of the convolutional encoders, called Recursive Systematic Convolutional (RSC) encoders, as illustrated by the general structure in Fig. 4.22. The output stream is built up by multiplexing a_i, $c_i^{(1)}$ and $c_i^{(2)}$ at each cycle i of the clock. For this reason, another classification of the convolution codes has to be discussed that differs from the one given above.

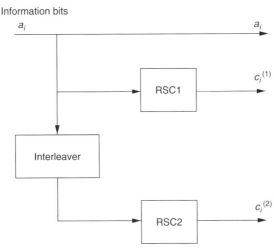

Figure 4.22 General structure of a turbo encoder

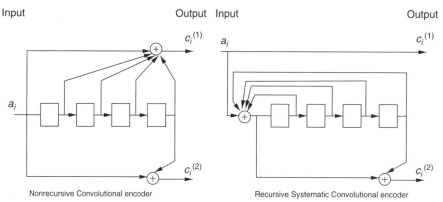

Figure 4.23 General structure of NRC and RSC encoders

Convolutional encoders can be categorized into two main categories: the traditional Non-Recursive Convolutional (NRC) encoders and the Recursive Systematic Convolutional (RSC) encoders. Figure 4.23 illustrates the structure of both of these encoders. The central component of the NRC encoder is the shift register, which stores previous values of the input stream. The output is then formed by linear combinations of the current and past input values. This encoder is nonsystematic; this means that the systematic (or input) data is not directly sent as an output. In contrast to this, the NRC encoders can be either systematic or nonsystematic. The figure also shows the structure of the RSC encoders that are commonly used in turbo codes. The RSC encoder contains a systematic output and a feedback loop, which is the necessary condition for the RSC realization.

The traditional turbo code encoder is built by concatenating two RSC encoders with an interleaver in between, as illustrated in Fig. 4.24 [Bing02]. Usually, the systematic output of the second RSC encoder is omitted to increase the code rate. Several performance investigations about the turbo codes were achieved in the last years and some of them are recommended for further information about the theory, the complexity reduction and design of these encoders and their decoders, especially in [Bing02, Li02, Garo03].

4.3.3 Interleaving

A common method to reduce the "burstiness" of the channel error is the interleaving, which can be applied to single bits or symbols to a given number of bits. Interleaving is the procedure which orders the symbols in a different way before transmitting them over the physical medium. At the receiver side, where the symbols are de-interleaved, if an error burst has occurred during the transmission, the subsequent erroneous symbols will be spread out over several code words. This scenario is illustrated in Fig. 4.25, showing a simple interleaving procedure where the elements of the original symbols 1 and 2 are interleaved element per element to build up two new symbols that will be transmitted over the channel. Suffering from disturbances, two adjacent elements of the transmitted symbol are destroyed, building a burst with the length of two elements. In the receiver, the received symbols are de-interleaved, and therefore the error burst is decomposed into two single element errors.

In the design of turbo encoders, the output code words of a RSC encoder have a high Hamming weight. However, some input sequences can cause the RSC encoder to

Realization of PLC Access Systems

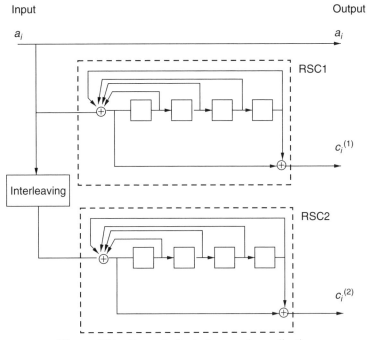

Figure 4.24 Example for turbo encoder realization

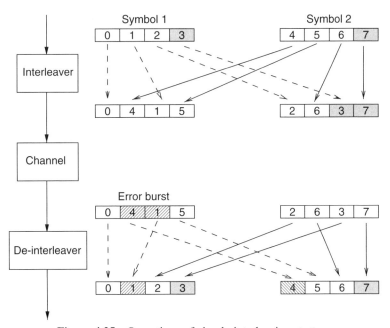

Figure 4.25 Operations of simple interleaving strategy

produce low weight code words. The combination of interleaving (permuting) and RSC encoding ensures that the code words produced by a turbo encoder have a high Hamming weight [Garo03]. For instance, assume that the RSC1 encoder implemented in the turbo encoder in Fig. 4.24 receives an input sequence that causes it to generate a low weight output. Then it is improbable that the other convolutional encoder RSC2, which receives the interleaved version of the input, will also produce a low weight output. Hence, the interleaver spreads the low weight input sequences, so that the resulting code words have a high Hamming weight.

Furthermore, in a statistical sense the interleaving might be interpreted as reduction of the channel memory, and a perfectly interleaved channel will have the same properties as the memoryless channel [Schu99]. The application of interleaving is limited by the added delay, because at the receiver side, the de-interleaver has to wait for all interleaved code words to arrive. This effect is not desired in the real-time applications. Different types of interleavers were developed over the last years, and this development is accelerated by their application within the turbo encoders.

Block interleavers accept code words in blocks and perform identical permutations over each block. Typically, block interleavers write the incoming symbols by columns to a matrix with N rows and B columns. If the matrix is completely full, the symbols are then read out row by row for the transmission. These interleavers are pointed out as (B, N) block interleavers. At the receiver side, the de-interleavers complete the inverse operation, and for this the exact start of an interleaving block has to be known, making the synchronization necessary. Properties of an interleaver block are

- any burst of errors of length $b \leq B$ results in single errors at the receiver, where each is separated by at least N symbols, and
- the introduced delay is of $2NB$, and the memory requirement is NB symbols, at both transmitter and receiver sides.

Unlike the block interleavers, the convolutional interleavers have no fixed block structure, but they perform a periodic permutation over a semifinite sequence of coded symbols. The symbols are shifted sequentially into a bank of B registers of increasing lengths. A commutator switches to a new register for each new code symbol, while the oldest symbol in that register is shifted out for the transmission. The structure of a convolutional interleaver is given in Fig. 4.26 [Schu99].

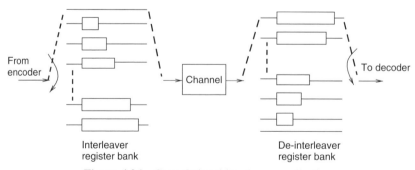

Figure 4.26 Convolutional interleaver realization

With apparition of turbo codes, the interleaving became an integral part of the coding and decoding scheme itself. One problem with classical interleavers is that they are usually designed to provide a specific interleaving depth. This is useful only if each burst of errors never exceeds the interleaver depth, but it is wasteful if the interleaver is over-designed (too long) and error bursts are typically much shorter than the interleaver depth [CrozLo99]. Furthermore, in practice, most of the channels generate usually error events of random length, and the average length can be time varying, as well as unknown. This makes it very difficult to design optimum interleaving strategies using the classical approaches. What is needed is an interleaving strategy that is good for any error burst length. Such strategy is proposed in [CrozLo99] and is called Golden Interleaving Strategy, which is based on a standard problem in mathematics called the "Golden Section". The design and implementation complexity of this strategy, which demonstrated higher performances in comparison to the other ones, are discussed in [Croz99].

4.3.4 ARQ Mechanisms

ARQ provides a signaling procedure between a transmitter and a receiver. The receiver confirms a data unit by a positive acknowledgement (ACK), if it is received without errors. A request for the retransmission of a data unit can be carried out by the receiver with a negative acknowledgement (NAK), in the case in which the data unit is not correctly received, or is missing. An acknowledgement is transmitted over a so-called reverse channel, which is also used for data transmission in the opposite direction. Usually, an acknowledgement is transmitted together with the data units carrying the payload information.

There are the following three basic variants of ARQ mechanisms [Walke99]:

- *Send-and-Wait* – Every data unit has to be confirmed by an ACK before the next data unit can be transmitted. The data unit has to be retransmitted if an NAK is received.
- *Go-back-N* – After the receiver has signaled that a data unit is disturbed, the sender has to retransmit all data units that are not yet acknowledged.
- *Selective-Reject* – After an NAK is received, the sender retransmits only a disturbed data unit. All correctly received succeeding data units do not have to be retransmitted.

The correctness of a data unit is proved on the receiver side by the usage of a CRC checksum in every data unit. To ensure realization of the ARQ mechanisms, transmitted data units have to be numbered with so-called sequence numbers. Thus, the order of the data units can be always controlled by the transmitting and receiving network stations.

4.3.4.1 Send-and-Wait ARQ

In accordance with the Send-and-Wait ARQ mechanism, after a transmitter sent a data unit (e.g. data unit No. 1, Fig. 4.27) it waits for an acknowledgement before it sends a next data unit. If the received acknowledgement was positive (ACK), the transmitter proceeds with transmission of the next data unit (provided with next sequence number). On the other hand, if the acknowledgement was negative (NAK), the transmitter repeats the same data unit.

It can be recognized that this variant of ARQ mechanisms is not effective. Especially, in the case of long propagation delays and small data units, data throughput seems to

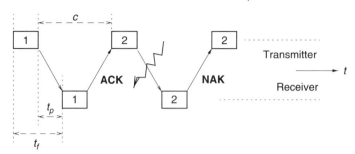

Figure 4.27 Send-and-wait ARQ

be low. The data throughput (S) for the Send-and-Wait mechanism can be calculated according to the following equation:

$$S = \frac{n \cdot (1 - DER)}{n + c \cdot v} \qquad (4.52)$$

n –   length of a data unit, in bits
DER – Error Ratio of Data units
c –   delay between end of transmission of last data unit and start of next data unit
v –   transmission rate

It is also possible that a data unit never arrives at the receiver (e.g. it is lost because of hard disturbance conditions). In this case, the transmitter would wait for an infinite time for either a positive or a negative acknowledgement to transmit the next data unit or to repeat the same one. To avoid this situation, a timer is provided within the transmitter to initiate a retransmission without receiving any acknowledgement. So, if the receiver does not receive a data unit and accordingly does not react with either ACK or NAK, the transmitter will retransmit the data unit after a defined time-out.

4.3.4.2 Go-back-N Mechanism

As is mentioned above, the limitation of the Send-and-Wait ARQ protocol are possibly long transmission gaps between two adjacent data units (data units with adjacent sequence numbers). To improve the weak data throughput, Go-back-N ARQ mechanism provides transmission and acknowledgement of multiple data units ensuring a near to continuous data flow between the transmitter and receiver. Thus, a transmitter can send a number of data units one after the other and receives an acknowledgement for the number of sent data units. According to the Go-back-N principle, the transmitter sends the data units without waiting for the acknowledgement from the receiver (Fig. 4.28). The maximum number of data units which can be sent without confirmation is specified by a so-called transmission window. After the transmitted units arrive, the receiver sends acknowledgement for all received data units.

If the transmitter receives a positive acknowledgement (ACK), all data units with sequence numbers less or equal to the acknowledged data unit are considered as correctly

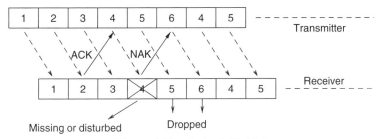

Figure 4.28 Go-back-N ARQ

transmitted. Afterward, it can proceed with transmission of further data units. The acknowledgement from the receiver can be also sent before it receives all data units from the transmit window (controlled by a receive window). So, the transmitter can proceed with the transmission before all data units of a transmit window are delivered to the receiver. If the transmitter receives a negative acknowledgement (NAK) for a data unit, it has to repeat this and all data units with higher sequence numbers. For this reason, the transmitter requires a sufficient buffer to keep all data units until they are acknowledged by the receiver. To keep the buffer requirement finite, the transmit window has to be limited as well.

To explain the Go-back-N ARQ, we consider an example presented in Fig. 4.28. In accordance with the Go-back-N mechanism, the sender transmits its data units, which are marked with a sequence number, continuously, until it receives an NAK signal from the receiver (e.g. for data unit with sequence number 4). After that, the transmitter again sends the requested data unit (4) and continues with the sending of all succeeding data units (5, 6, ...). The receiver ignores all data units which are not in-sequence until the next in-sequence data unit (4) arrives. Afterward, the receiver accepts all succeeding data units too (5, 6, ...). If the receiver sends a positive acknowledgement (ACK), it confirms the correctly received data unit (e.g. data unit 2) and also all data units with the lower sequence numbers.

By the usage of Go-back-N mechanism, data throughput S is improved compared with Send-and-Wait mechanism, and can be calculated according to the following equation [Walke99]:

$$S = \frac{n \cdot (1 - DER)}{n + DER \cdot c \cdot v} \qquad (4.53)$$

n – length of a data unit, in bits
DER – Error Ratio of Data units
c – delay between end of transmission of the last data unit and start of the next data unit
v – transmission rate, in bps

4.3.4.3 Selective-Reject

A further improvement of the ARQ efficiency is ensured by the Selective-Reject mechanism. In this case, the NAK's are sent for data units that are missing or disturbed, such as in Go-back-N mechanism. However, opposite to the Go-back-N ARQ, the transmitter repeats only with NAK requested data units. Other data units with higher sequence number are considered as correctly received (of course if there is no NAK for these data units) and they are not retransmitted. Thus, the Selective-Reject mechanism achieves better data throughput, as expressed by the following equation, if it is assumed that the receiving storage capacity is unlimited [Walke99]:

$$S = 1 - DER \qquad (4.54)$$

For the realization of the Selective-Reject mechanism, it is necessary that the receiver buffer is large enough to store the data units until the data units with lower sequence numbers arrive at the receiver. The transmitter can remove data units form the buffer after it receives an acknowledgement, such as in the Go-back-N mechanism.

4.4 PLC Services

As is mentioned in Sec. 4.1, the aim of the considerations carried out in this chapter is a description of the PLC specific protocol stack (Fig. 4.2). So, we considered suitable modulation schemes for PLC in Sec. 4.2 and various error handling mechanisms in Sec. 4.3. The MAC sublayer to be applied in PLC networks is separately investigated in Chapters 5 and 6. On the other hand, a PLC network is used for realization of various telecommunications services. Thus, the PLC specific protocol stack, specified within so-called PLC-specific network layers in Sec. 4.1, has to be able to ensure realization of these services. Accordingly, various services causing different data patterns can be considered as an input for the PLC-specific network layers. For this reason, in this section, we analyze telecommunications services that are expected to be used in PLC networks and discuss their traffic characteristics. This allows specification of different source models to be applied in investigations of PLC networks and specification of requirements on the PLC-specific protocol stack to support realization of various services.

4.4.1 PLC Bearer Service

An access network provides transport bearer capabilities for the provision of telecommunications services between a service node and subscribers of the access network [MaedaFe01]. Accordingly, a PLC access network can also be considered as a bearer service, providing telecommunication services to the subscribers within one or multiple low-voltage power supply networks. A bearer service (or a bearer/transport network), such as classical telephony network, X.25 packet network, ATM network, and so on, carries teleservices, which allow usage of various communications applications (Fig. 4.29). According to the functions of bearer services, to provide transport capabilities for various telecommunications services, they are specified within so-called network layers of the ISO/OSI reference model (Sec. 4.1).

As mentioned in Sec. 2.3.4, PLC access networks cover only the last part of an entire communications path between two subscribers. The entire communications path consists of access and distribution networks, as well as backbone network and is probably realized by a number of different communications technologies. Thus, a PLC access network provides bearer service only for a certain part of the communications path. Therefore, PLC networks have to be able to exchange information with other communications systems

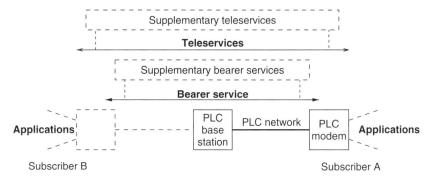

Figure 4.29 Classification of telecommunications services

that are offering bearer services as well (e.g. technologies used in distribution networks for connection to the backbone). In other words, PLC has to be compatible with other communication technologies, in so far that interconnection between various systems is possible, which is ensured by compatibility of different bearer services.

Teleservices cover the entire telecommunications functions including all communications layers (1 to 7) specified in the ISO/OSI reference model (Fig. 4.1). Accordingly, the teleservices functions are implemented in subscriber's communications devices (Fig. 4.2) and they are not included within PLC-specific protocol stack. However, PLC networks have to provide capabilities for realization of various teleservices, such as telephony, internet access, and so on. The basic function of both bearer services and teleservices (Fig. 4.29) can be extended by different additional features, building so-called supplementary services [itu-t93]. Thus, basic telephony teleservices can be extended to include various features, such as wake- up service, number identification service, and many other services that are offered in modern telecommunications networks.

From the subscriber's point of view, the teleservices are used for realization of various communications applications (Fig. 4.29). So, telephony is used for speech, as a communications application, and internet access can be used for realization of numerous data applications, such as WWW browsing, messenger services, internet games, and so on. The subscribers (users of telecommunications services) judge a network, a service, or a network provider in accordance with the quality of communications applications they use. On the other hand, PLC networks have to offer a large palette of telecommunications services with a satisfactory QoS, to be able to compete with other communications technologies applied to the access networks (Sec. 2.1). Therefore, PLC access networks have to provide a bearer service that can carry different teleservices, ensuring various communications applications. Accordingly, the entire protocol stack (Sec. 4.1) to be implemented in PLC networks has to provide features to allow transmission of different kinds of communications information produced by various teleservices and applications. At the same time, it is important to ensure certain QoS in PLC access network, as well.

4.4.2 Telecommunications Services in PLC Access Networks

As concluded above, PLC networks have to offer various telecommunications services to be able to compete with other access technologies, to attract possibly higher number of PLC subscribers, and to ensure economic efficiency of the PLC networks. Therefore, PLC networks must support the classical telephone service because of its importance and huge penetration in the communications world. Telephony is still the most acceptable communications application, requiring relatively simple communications devices and low technical knowledge for customers using this service. Furthermore, in spite of a rapid development of various data services in the last decades, network operators still achieve large revenues by offering the telephony service.

On the other hand, another important telecommunications service is data transmission allowing broadband internet access. Nowadays, we can observe a rapid development of various communications applications based on the internet service in business, as well as in private environment. In accordance with the current acceptance of internet applications, we can expect that in the near future internet access will be more and more spread in the communications world, similar to the case with the telephony service. Therefore, both telephony and internet services are considered as primary telecommunications services that have to be realized by broadband PLC access networks.

4.4.2.1 Telephony

In the classical telephony service, a certain portion of the network capacity (e.g. 64 kbps) is allocated for a voice connection for its entire duration. The voice connections are characterized by two parameters: interarrival time between calls and holding time [Chan00]. Generation of new calls is considered as a Poisson arrival process. Accordingly, in traffic models representing the classical telephony service, the interarrival time of the calls as well as the holding time (duration of the calls) can be represented by random variables that are negative exponentially distributed (Fig. 4.30).

The Probability Distribution Function (PDF) for the interarrival time is expressed as

$$A(t) = 1 - e^{-\lambda \cdot t}, t \geq 0 \qquad (4.55),$$

representing probability that no arrivals occur in interval (0, t) [Klein75]. Its probability density function (pdf) is

$$a(t) = \lambda e^{-\lambda \cdot t}, t \geq 0 \qquad (4.56),$$

where λ is arrival rate of calls. The mean for the exponential distribution is calculated as $1/\lambda$, in this case representing mean interarrival time of calls.

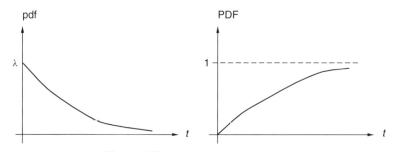

Figure 4.30 Exponential distribution

Of course, Eq. (4.55) and Eq. (4.56) are also used for representation of negative exponentially distributed holding time (duration of the calls), where $\frac{1}{\mu'}$ is mean holding time of a call.

Negative exponential distribution, described above, is widely applied in performance evaluations regarding the classical telephony service. The mean duration of calls is considered to be between two and three minutes.

The speech as a communications application is not continuous and consists of so-called talkspurt and silent periods (Fig. 4.31). This is caused by the nature of a conversation, where the speakers make pauses between words, sentences, and also by the periods when a conversation participant listens to another. Since for a telephony connection a certain network capacity is allocated for its entire duration, the allocated network capacity is not used during the silent periods, which is not efficient. Therefore, methods for usage of the silent periods of the telephony connections had already been applied to the transport networks decades ago to improve efficiency of, at that time limited and expensive, intercontinental links. So-called packet voice/telephony service is also considered for applications in different wireless communications systems, to improve utilization of still limited data rates in these networks. For the same reason, that is, the efficient use of the limited network capacity, application of the packet voice service is also of interest in PLC access networks.

If the packet voice service is applied, the speech information is transmitted only during the talk periods and the silent periods can be used by other connections and services. In this way, either data or speech information from another packet voice connection can be transmitted over the same link. During the talkspurts of a voice connection, the speech information is transmitted in special data packets. The packet voice connections are characterized by two parameters: duration of talkspurts, used for transmission of a number of the voice packets, and duration of silence periods. These parameters are represented by two corresponding random variables, specified in the appropriate traffic models, which are considered for application in different communications technologies (e.g. [LiuWu00, LenzLu01, FrigLe01a]), but can also be applied for investigations of the PLC networks. However, the interarrival time and the entire duration of packet voice connections can be modeled in the same way, such as in the case for the classical telephony service, by usage of the traffic models specified above.

It can be intuitively recognized that the usage of silent periods in packet voice service improves the network efficiency, compared with the classical telephony service. However,

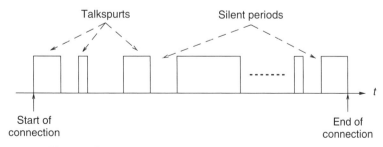

Figure 4.31 Busy and silent phases of a voice connection

in the case of packet voice the continuous information flow provided in the classical telephony does not exist and the voice packets can be delayed, especially if the number of subscribers using the same transmission medium increases. Since the voice connections are very sensitive to larger delays, making a telephony conservation less understandable or even not possible, there are some limits for the maximum delay of the voice packets. For example, for wireless networks applied in the access area, the maximum delays are set to 20–24 ms ([AlonAg00, KoutPa01]), or to 25 ms to avoid the usage of echo cancellers [DaviBe96]. Note that, access networks, such as PLC, cover only a part of the common transmission path between the participants of a voice connection. Therefore, the delay limits for the telephony service in the access area are stronger than for the entire transmission path.

In recent telecommunications networks, different kinds of services are transmitted over the same transmission links, with a possible usage of same networks elements, such as switching devices, routers, and so on. Nowadays, telecommunications networks worldwide are based on the internet protocol (IP), originally developed for pure data transmission and not designed for transmission of the voice. However, due to the trends for integration of voice and data services, a solution for the realization of the voice service over IP networks, so-called Voice over IP (VoIP), is seriously considered as a solution for so-called integrated services networks. Therefore, VoIP is considered for applications in broadband PLC access networks as well.

Recent experience with the VoIP service shows that the QoS requirements on the voice service (e.g. delay, losses, etc.) can be well met in low-loaded networks. However, if an IP network is highly loaded, the performance regarding VoIP decreases significantly. Therefore, various mechanisms for the traffic control are considered to be applied in the high-speed networks, ensuring a required QoS for the time-critical services also, such as voice. On the other hand, the available data rates in PLC access networks are significantly smaller than in the high-speed transport networks. Thus, the traffic control mechanisms developed for providing data rates beyond 100 Mbps cannot be sufficient to ensure both sufficient QoS for the voice service and a good network utilization in networks with limited data rates (few Mbps). Therefore, it is necessary to provide additional mechanisms within the PLC protocol stack, providing required QoS for the voice and other critical telecommunications services and simultaneously ensuring a good network utilization.

4.4.2.2 Internet Access

The most used telecommunications service in recent PLC access network is data transmission based on the internet access. Therefore, it is necessary to analyze internet data traffic and outline the main characteristics of such traffic patterns, which are typical for the access networks, such as broadband PLC networks. The traffic characterization (see e.g. [FärbBo98]) is carried out with the help of numerous measurements in different networks, to achieve possibly more general results and to allow design of appropriate models, representing possibly real traffic characteristics. However, during the last decade, the traffic characteristics have been frequently changed, because of the rapid development of new telecommunications services, a very intensive growth of wireline and wireless networks, increase in the number of subscribers and operators, and so on. Accordingly, the changing nature of the traffic characteristics is also recognized in many research studies (e.g. [SahiTe99, FeldGi]). Therefore, the traffic models cannot represent an exact traffic characteristic and its future variations, but they can be chosen to represent a generalized traffic behavior, according to a specific investigation aim, in this case, a performance evaluation of PLC access networks and specific PLC protocol stack.

The application mainly used in internet is World Wide Web (WWW). Accordingly, most traffic models representing behavior of internet users are developed according to WWW traffic patterns. The characterization and modelling of the WWW traffic can be done on the following levels, as proposed in [ReyesGo99]:

- Session level, representing a working session of a user with a WWW browser from the time of starting the browser to the end of the navigation,
- Page level, including visits to a WWW page and considering a page as a set of files (HTML, images, sounds, etc.), and
- Packet level, the lowest level representing transmission of IP packets.

A session is defined as the work of an internet user with a WWW browser. It includes the download of a number of WWW pages and viewing of the pages (Fig. 4.32). Generally, a WWW page consists of a number of objects (different files, images, etc.) that are simultaneously transmitted during a page download. A first request for a WWW page, which is manually carried out by an internet user, causes a download of a main page object [TranSt01]. The main object is followed by a number of so-called in-line objects, which are automatically requested by a browser or just transmitted by an internet server as a logical succession to the main object.

The transmission of each page object causes the establishment of a separate TCP connection. During a TCP connection [Stev94], there is a data exchange between a transmitter (e.g. internet server) and a receiver (e.g. internet user), including a transmission of user data and various control messages, such as TCP acknowledgements. The transmitted data units on the TCP level correspond to the IP packets (a TCP packet contains an IP packet plus TCP overhead). To be transmitted over a network, the IP packets are delivered from

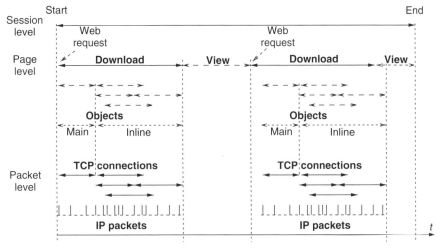

Figure 4.32 Internet user behavior on different observation levels

the upper network layers to the data link layer (Fig. 4.1). From the point of view of the data link layer, the IP packets can be considered as input data units. Accordingly, IP traffic models are suitable to represent the data traffic, such as WWW, in the investigation of the PLC specific protocol stack (Fig. 4.2). A detailed description of traffic models that are used for performance evaluation of PLC MAC layer representing WWW-based internet traffic is presented in Sec. 6.2.3.

4.4.2.3 Advanced Broadband Services

A further requirement on the broadband PLC access networks and its development is to offer so-called advanced broadband services. Thus, beside primary telecommunications services described above, the future PLC systems have to offer services using higher data rates with higher QoS requirements (priorities, delays, etc.), such as video. However, recent PLC networks allow data rates up to several Mbps over a shared transmission medium, which is not sufficient for realization of the services with higher data rates, at least in the case that a higher number of subscribers using such services are connected to a PLC access network.

On the other hand, a rapid development of various communications technologies, including new transmission methods, modulation schemes, and so on, can also speed up the development of PLC systems with higher data rates in the near future. Therefore, the PLC protocol stack has to be flexibly designed to allow realization of different variations of QoS guarantees required by both recent as well as future telecommunications services and applications. However, in the first place, the usage of the telephony and internet access (primary services) has to be realized with the required QoS to ensure an initial impact of the PLC systems in the competition with other access technologies.

4.4.2.4 Narrowband Services

In Sec. 2.2.4, we considered narrowband PLC systems, ensuring realization of various so-called specific PLC services, such as home automation, energy management, various security functions, and so on. In this case, various devices using electrical power can be easily connected over the same grid to a PLC system, which can be used for the remote control of such devices. Narrowband PLC systems are already standardized and they are also widely available for usage by both business and private consumers. However, integration of the narrowband PLC services into broadband PLC networks would improve the initial position of PLC systems on the market compared with other communications technologies. Therefore, the integration of both narrowband and broadband PLC systems should be seriously considered during the design of broadband PLC networks.

The PLC-specific services are supposed to use significantly lower data rates than telephony, internet and other typical telecommunications services and they usually do not require high QoS guarantees. Therefore, the realization of the narrowband services and their integration within broadband PLC systems seem not to be critical. On the other hand, some specific narrowband services can require very low response time (delays) in the case when they ensure transmission of some significantly important information (e.g. temperature alarms, security messages, etc.). Of course, integrated narrowband–broadband PLC systems have to be able to fulfill such specific QoS requirements. However, if the requirements of time-critical telecommunications services, such as voice, can be met by a broadband PLC network, the realization of the critical narrowband services should be not critical as well.

4.4.3 Service Classification

In the previous section, we considered several telecommunications services, that are expected to be used in access networks, such as PLC. However, telecommunications are one of the most growing technological areas in this era with a rapid development of new services and applications. It can also be observed that telecommunications services continuously change their nature in accordance with the development of new communications technologies and applications. This causes changes of the traffic patterns transmitted over the communications networks, as well as significant variations of required QoS guarantees for different kinds of services.

Because of the increasing number of services with different features and requirements, it is not possible to consider all possible services during the design of various communications networks and transmission systems. Additionally, it is also not possible to take into account telecommunications services that do not exist at present and will be developed in the future. Therefore, it is necessary to classify the services according to their main features represented by the QoS requirements and traffic characteristics. A first classification of telecommunications services including both primary services can be done according to their nature, as listed below:

- Circuit switched services (e.g. telephony), and
- Packet switched services (data transfer without QoS guarantees, e.g. internet access).

However, recent development of telecommunications services calls for a finer service classification. In the consideration below, we present service classification used in several recently applied telecommunications networks, which can be used in the investigations of broadband PLC access networks too.

4.4.3.1 Traffic Classes

In specifications for UMTS networks, telecommunications services are divided into four groups: conversational, streaming, interactive and background, as is presented in Tab. 4.3, [QiuCh00]. The services are classified according to the traffic characteristics caused by different communications applications that are expected in modern telecommunications networks. Each traffic class has specific QoS requirements depending on the nature of the used applications.

Thus, for a typical application belonging to the conversational traffic class, such as voice, it is important to ensure very low delays, as is also mentioned in Sec. 4.4.2. In the case of the streaming traffic class (e.g. video), a certain time relation between transmitted streaming packets/frames has to be ensured (e.g. to have a near to continuous video stream ensuring a satisfactory signal reception on the target device). The same requirement is necessary for the conversational traffic class if packet voice (or VoIP) service is applied.

Table 4.3 UMTS traffic classes

Traffic class	Conversational	Streaming	Interactive	Background
Characteristics	– Preserve time relation between information entities of the stream – Conversational pattern (stringent and low delay)	– Preserve time relation between information entities of the stream	– Request response pattern – Preserve payload content	– Destination is not expecting the data within a certain time – Preserve payload content
Application	Voice	Streaming video	Web browsing	E-mail

The most used application belonging to the interactive traffic class is web browsing (Sec. 4.4.2). In this case, the time relation between transmitted packets is not important and the delays are not so critical. However, an interactive web user expects a response from a remote server within a reasonable time interval. The background traffic class has the lowest delay requirements and includes applications that can be served by a network with a lower priority. E-mail or file transfer without delay requirements are typical representatives of such telecommunications applications.

4.4.3.2 Service Categories

As mentioned in Sec. 4.4.1, a PLC access network considered as a bearer service has to ensure transmission of different teleservices providing different communications applications. Accordingly, specific PLC protocol stack has to provide several bearer service categories to carry information arising from different traffic classes, described above.

To provide various telecommunications services with different traffic characteristics (e.g. as classified by UMTS traffic classes), integrated services networks have to ensure a simultaneous transmission of various traffic patterns with different QoS requirements. For this purpose, the integrated services networks provide so-called service categories, which are specified to ensure transmission of varying types of services with similar traffic characteristics and QoS requirements. A service categorization is done for the integrated services in internet, specifying the following three classes [ConnRyu99]: guaranteed service (GS), controlled load (CL), and best effort services.

- GS category is designed to meet the QoS requirements of real-time services, such as voice and video, with very strong delay limits and very low packet loss. To provide such types of services, a certain network capacity can be strictly allocated a real-time connection for the entire time of its duration, or a connection admission control (CAC) mechanism has to be implemented to ensure the required QoS without fixed allocation of the transmission resources, or a combination of both methods can be applied.
- CL category is provided for connections and services that can tolerate higher packet loss and longer delays than in GS category. A fixed capacity allocation is not provided by the CL category, and accordingly, network resources can be used efficiently.

- Best Effort category provides no QoS guarantees (e.g. delay) and the flow control is placed at the transport layer (is not a network function, see Sec. 4.1).

In the context of ATM networks, as a second example of the integrated service networks, the following service categories are specified [ConnRyu99]:

- CBR – constant bit rate service, which is accomplished by allocating a fixed amount of network resources for the entire duration of a connection,
- VBR – variable bit rate service, requiring strict boundaries on delay, delay variation and packet loss, which is divided in two subcategories:
  - rtVBR – real-time VBR with the same characteristics as GS internet service category (described above), and
  - nrtVBR – non-real-time VBR without delay limitations, but with the requirement for low packet loss
- ABR - available bit rate service, which gives QoS guarantees that can possibly change over the life of the connection and is similar to CL internet service category, and
- UBR - unspecified bit rate service without any QoS guarantees, which is the same as best effort category specified for internet.

4.5 Summary

PLC network elements, such as modem and base station, make possible information exchange between various communications devices over electrical supply networks. A PLC modem has usually several interfaces for connection with different end user systems, whereas a PLC base station provides interfaces for interconnection through the backbone network. All these functions of the PLC network elements are specified in network layers of ISO/OSI reference model. A particular function of a PLC network and its elements is communication over power grids and coupling of the communications devices to the electrical installation. This is ensured by particular coupling, transmission and communications methods, as well as access protocols and error handling mechanisms, specified in first two network layers, physical and data link layer, which build a specific PLC protocol stack.

The modulation technique to be applied for PLC physical layer has to overcome the strong PLC channel impairments in order to realize high bit rates. Furthermore, the modulation scheme has to realize acceptable BER with a SNR as low as possible, to be able to coexist with other systems already deployed in its environment and to guarantee a satisfactory quality of service. Two main modulation schemes are the subjects of investigations and trials, namely the Orthogonal Frequency Division Multiplex (OFDM) and the Spread Spectrum, with its two forms Direct Sequence and Frequency Hopping. These solutions have already shown very good performances, and therefore are already standardized for widely deployed systems, such as the Asymmetrical Digital Subscriber Line (ADSL) and Digital Audio Broadcasting for OFDM, and the Wireless Local Access Networks (WLAN) for the DSSS. However, each of these candidates has advantages and drawbacks and a kind of trade-off has to be managed in order to meet the aimed performances.

Error handling in PLC networks is carried out on different levels of the protocol stack. Thus, a sufficient SNR ensures robustness of PLC against background noise. On the other hand, specific transmission methods used for broadband PLC can partly avoid influence of impulsive and narrowband disturbances. As with many other communications systems, PLC networks have to apply FEC and interleaving mechanisms to detect, and correct in the normal case, the transmission errors. Several FEC techniques can be used, such as block codes, convolution codes and the turbo codes, which consists of a stronger class of error correcting codes. However, the selected coding scheme has to guarantee a reasonable complexity of the encoder/decoder and to not introduce more delay. Because the performances of the these codes are strongly limited by the error bursts, interleavers became an integrated part of each encoder. Therefore, besides the classical interleaving strategies, the block and convolutional interleaving, new ones have appeared in the last years, such as the Golden interleavers and their derivatives. The erroneous data units that are not corrected by FEC are retransmitted by an ARQ mechanism. To reduce the influence of long-term disturbances, PLC systems have to be able to reallocate transmission resources ensuring continuous network operation.

A PLC access network is considered as a bearer telecommunications service carrying various teleservices, which are used by different communications applications. To be able to compete with other access technologies, PLC has to ensure realization of a large pallet of telecommunications services with a sufficient QoS. Therefore, PLC access networks have to provide various bearer service categories allowing transmission of different traffic flows, caused by various telecommunications services. For the investigation of the PLC protocol stack, the different services are represented by appropriate source models that depict their traffic characteristics.

5

PLC MAC Layer

In this chapter, we consider various solutions for the MAC layer to be applied to broadband PLC access networks. Components of a MAC layer and requirements on a suitable realization for the PLC MAC layer are presented in Sec. 5.1. Afterward, we describe several multiple access schemes and MAC protocols and analyze possibilities for their application in PLC networks. Finally, the traffic control mechanisms are discussed. A detailed investigation of reservation MAC protocols for the application in PLC is presented separately in Chapter 6.

5.1 Structure of the MAC Layer

5.1.1 MAC Layer Components

The basic task of a MAC layer is to control access of multiple subscribers connected to a communications network using a same, so-called "shared transmission medium", and organization of information flow from different users applying various telecommunications services. Generally, functions of a MAC layer applied to any telecommunications network can be divided into the following three groups:

- Multiple access
- Resource sharing
- Traffic control functions.

A multiple access scheme establishes a method of dividing the transmission resources into accessible sections [AkyiMc99], which can be used by network stations for transmission of various information types. The choice for a multiple access scheme depends on the applied transmission system within the physical layer and its features. Following the definition of a multiple access scheme, there is a need for specification of the strategy for the resource-sharing MAC protocol. The task of a MAC protocol is the access organization of multiple subscribers using the same shared network resources, which is ensured by management of the accessible sections provided by the multiple access scheme. Duplex mode is one of the functions of the MAC layer controlling the traffic between downlink and uplink transmission directions. Additional traffic control functions, such as traffic scheduling, admission control, and so on, can be implemented in higher network layers, but also completely or partly within the MAC layer. In any case, to fulfill QoS requirements of various telecommunications services, MAC layer and its protocols have to be able to support realization of different procedures for traffic scheduling, as well as to support implementation of a Connection Admission Control (CAC) mechanism.

In this section, we describe characteristics of the MAC layer to be applied in broadband PLC access networks and specify technical requirements on the PLC MAC layer. In Sec. 5.2, we present possibilities for realization of multiple access schemes and their application in PLC networks. Afterward, we define a generalized channel model to be used for investigation of different multiple access schemes. In Sec. 5.3, we present various strategies for the resource sharing and analyze their application in PLC networks. A comprehensive investigation of reservation MAC protocols for the application in PLC is given in Chapter 6. In Sec. 5.4, we present solutions for the duplex mode and briefly discuss possibilities and requirements for realization of traffic scheduling and admission control in PLC access networks.

5.1.2 Characteristics of PLC MAC Layer

The MAC layer is a component of the common protocol architecture in every telecommunications system with a shared transmission medium. There are various realizations of the MAC layer and its protocols that are developed for particular communications networks, depending on their specific transmission features, operational environment, and their purpose. The particularity of PLC access networks includes a special transmission medium (low-voltage power supply network) providing limited data rates under presence of an inconvenient noise scenario causing disturbances for data transmission. On the other hand, to ensure the competitiveness with other access technologies, PLC has to offer a wide palette of telecommunications services and to provide a satisfactory QoS.

The following four factors have a direct impact on the PLC MAC layer and its protocols (Fig. 5.1): network topology, disturbance scenario, telecommunications services and applied transmission system.

Network topology of a PLC access network is given by topology of its low-voltage power supply network, which is used as a transmission medium, having a physical tree topology. However, for the investigation of higher network layers (above physical layer), such as the MAC layer, a PLC access network can be considered as a logical bus system (Sec. 3.1.5) with a number of network stations using the same transmission medium to communicate with the base station, which connects the PLC network to the WAN. Influence of different kinds of noise, causing disturbances in PLC networks, and various telecommunications services used in PLC access networks are represented in investigations of the MAC layer by appropriate models; disturbance and traffic models, considered in Sec. 3.4.4 and Sec. 4.4.2, respectively. In Sec. 4.2, we outlined two suitable solutions for realization of broadband PLC transmission systems; OFDM and spread-spectrum schemes with specific features have to be considered in development of the PLC MAC layer as well.

5.1.3 Requirements on the PLC MAC Layer

A multiple access scheme and a strategy for resource sharing (a MAC protocol) are situated in core of the MAC layer (Fig. 5.2). As is mentioned above, the multiple access scheme

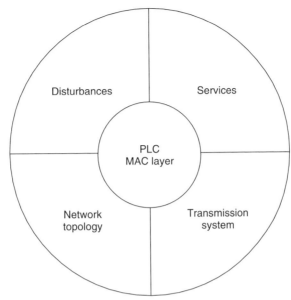

Figure 5.1 Environment of PLC MAC layer

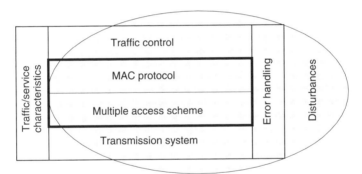

Figure 5.2 Structure of the MAC layer

establishes a method of dividing the transmission resources into accessible sections, and it depends on the applied transmission system within the physical layer and its features. In the PLC system under consideration, the multiple access scheme has to be applicable to the transmission system chosen for the PLC network (e.g. spread-spectrum or OFDM, Sec. 4.2). On the other hand, the task of the MAC protocol is the access organization of multiple subscribers using the same shared network resources, which is ensured by management of the accessible sections specified by the multiple access scheme. Accordingly, the MAC protocol has to be suitable for application on the multiple access scheme.

As is described in Sec. 3.4, PLC access networks operate under unfavorable noise conditions influencing the entire PLC protocol stack and causing disturbances for data transmission. Therefore, both the multiple access scheme and the MAC protocols have to be robust against the disturbance scenario, which is expected to exist in PLC networks. Furthermore, the unfavorable disturbance conditions cause application of various mechanisms for error handling in PLC networks. Accordingly, both multiple access scheme and MAC protocol have to be designed to allow integration of the error-handling mechanisms, such as ARQ (Sec. 4.3.4).

On the other hand, PLC access networks have to offer a number of telecommunications services and provide realization of QoS guarantees for different kinds of traffic classes (Sec. 4.4). Thus, both multiple access scheme and MAC protocols for PLC have to be adequate for realization of various QoS requirements for a traffic mix caused by different telecommunications services. The QoS provision is also ensured by application of additional traffic control mechanisms (Fig. 5.2), including duplex mode, traffic scheduling and admission control. However, the traffic control has to be designed to be robust against disturbances and to allow implementation of the error-handling mechanisms as well.

A further requirement on the PLC MAC layer is provision of a good network utilization, ensuring an economic efficiency of the PLC access networks. This can be ensured by an optimal management of available transmission resources provided by the multiple access scheme, carried out by the MAC protocol, as well as traffic control and error-handling mechanisms.

5.2 Multiple Access Scheme

As is mentioned above, a multiple access scheme establishes a method of dividing the transmission resources into accessible sections, which are used by multiple subscribers using various telecommunications services. A multiple access scheme is applied to a transmission medium (e.g. wireline or wireless channel) within a particular frequency spectrum, which can be used for information transfer. In the case of multiple subscribers using a shared transmission medium, telecommunications signals (information patterns) from individual users have to be transmitted within separated accessible sections, provided by a multiple access scheme, ensuring error-free communications. For this purpose, the signals from different subscribers, when they are transmitted over a shared medium, have to be orthogonal to each other, as presented by Eq. (5.1) [DaviBe96].

$$\int_{-\infty}^{\infty} x_i(t) x_j(t) \, dt = \begin{cases} 1 & i = j \\ 0 & \text{else} \end{cases} \quad (5.1)$$

In practice, it is not possible to achieve a perfect orthogonality between different signals using a same transmission medium. However, if influence between different signals is small enough, it can be accepted in communications systems. Generally, there are the following three multiple access schemes that can be applied in various communications systems:

- TDMA – Time Division Multiple Access,
- FDMA – Frequency Division Multiple Access, and
- CDMA – Code Division Multiple Access.

These three basic multiple access schemes can also be applied in various implementation combinations.

5.2.1 TDMA

5.2.1.1 Principle

In a TDMA scheme, the time axis is divided into so-called "time slots" that represent the accessible portions of transmission resources provided by the multiple access scheme (Fig. 5.3). Each time slot ensures transmission of a prespecified data unit, which can carry different kinds of information; speech sample, data packet, and so on. Usually, a time slot is used by only one user. In accordance with the TDMA, the data units, transferred within the time-slots, are transmitted by using the entire available frequency spectrum of the transmission medium.

The resource division in the time domain cannot be realized ideally without a separation of the time slots. Therefore, there are so-called "protection intervals" between time slots, ensuring that data from two neighboring time slots does not interfere. The time slots in a TDMA system can have a fixed or variable duration, allowing transmission of data units with a fixed specified size, or data units with variable sizes. In most TDMA systems, the time slots are organized in so-called "frames". Thus, a user with the permission to use a time slot can access exactly one slot with a precise position within a time frame.

TDMA is widely used in various communications networks. So, the modern telephone networks and cellular mobile networks, such as GSM, also apply TDMA principle. In these cases, a time slot is allocated for a telephony connection for its entire duration and repeats in every time frame. However, TDMA is also used for data transfer, where the time slots are usually dynamically allocated to different data connections. The TDMA scheme can be implemented in different transmission systems. Thus, in spread-spectrum and OFDM-based transmission systems, which are considered as suitable for realization of broadband PLC systems (Sec. 4.2), TDMA can be applied as well. In the case of an SS/TDMA (a combination of a spread-spectrum transmission system and TDMA scheme), a used frequency range in a network is divided into time slots, as is presented in Fig. 5.3. Features of a combination between OFDM and TDMA are described below.

5.2.1.2 OFDM/TDMA

As is described in Sec. 4.2.1, OFDM systems have a slotted nature where the transmitted information is divided into a number of OFDM symbols with a certain duration. Therefore, the application of TDMA schemes seems to be an appropriate solution for the

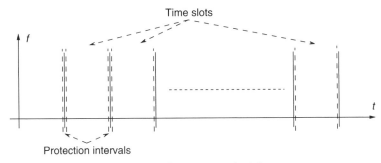

Figure 5.3 TDMA principle

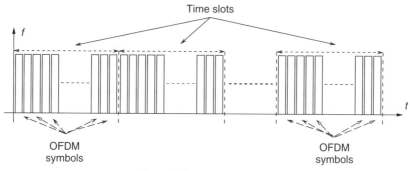

Figure 5.4 OFDM/TDMA

network based on the OFDM building an OFDM/TDMA transmission system [Lind99, WongCh99]. In this case, the network resources are divided into time slots, each of them carrying an integer number of OFDM symbols (Fig. 5.4). The length of the time slots can be fixed or variable, but the number of OFDM symbols within a time slot has to be an integer.

Some of the OFDM subcarriers can fail because of the disturbances (e.g. because of the long-term narrowband noise, Sec. 3.4), or they can operate with variable data rates if bit loading is applied. In both cases, the entire network capacity changes dynamically, according to the actual disturbance conditions. An OFDM symbol includes a particular number of bits/bytes and carries a specific amount of user data payload. Thus, if the network capacity is decreased, the payload of an OFDM symbol is reduced as well.

There are the following two solutions to keep the payload of an OFDM symbol constant:

- There are a number of so-called "spare subcarriers" that can be used in the case of failures or capacity decrease. However, if the disturbance conditions are more convenient at the moment, the spare subcarriers remains unused, which is not efficient.
- The duration of OFDM symbols is dynamically changed according to the current network capacity and availability of the subcarriers. Thus, the duration of the OFDM symbols is varied so that an OFDM symbol always carries a fixed amount of payload bytes. However, after each capacity change, the system has to be again synchronized to adapt to the lengths of the time slots and to fit an integer number of OFDM symbols.

To avoid the change of both symbol and time slot duration, the size of user data transmitted within a time slot can be variable to fit within an OFDM symbol, according to the actual network conditions and its currently available transmission capacity.

5.2.1.3 Data Segmentation

The division of the transmission resources in the time domain usually causes segmentation of larger data units (e.g. IP packets) into smaller data units. This is necessary because the data has to fit into data segments carried by the time slots provided by a TDMA scheme. At the same time, the data segmentation ensures a finer granularity of the network capacity and a simpler realization of QoS guarantees. Thus, if network resources are divided into smaller accessible portions, it is easier to manage the network resources and share them between various telecommunications services, ensuring realization of their particular QoS requirements. Furthermore, the data segmentation also ensures a higher efficiency in the case of disturbances. So, if a disturbance occurs, a data segment or a number of segments is damaged, and only damaged segments should be retransmitted (e.g. by an ARQ mechanism,). Accordingly, a smaller portion of the network capacity is used for the retransmission, which improves the network utilization.

On the other hand, a data segment consists in a general case of two parts; a header field and a payload field. The payload is used for storage of the user information to be transmitted over the network, and the header field consists of information needed for the control functions of the MAC and other network layers (e.g. control of data order, addressing, etc.). Therefore, the segmentation causes an additional overhead and there is a need for optimization of the data segment size, which depends on the disturbance characteristics in network.

An optimal segment size can be chosen in accordance with the BER in a communications system, as is presented in [Modi99]. If a network applying a perfect retransmission algorithm is considered, such as selective-reject ARQ (Sec. 4.3.4), the optimal segment size to be used in the network can be calculated according to the Eq. (5.2).

$$S_{opt} = \frac{-h\ln(1-p) - \sqrt{-4h\ln(1-p) + h^2\ln(1-p^2)}}{2\ln(1-p)} \quad (5.2)$$

p – channel bit-error-rate
h – number of overhead bits per segment

Figure 5.5 shows the optimal segment size, depending on the BER in a network, calculated for $h = 40$ overhead bits (5 bytes) per segment. With an increasing BER, segments errors become more frequent, and accordingly it is often necessary to retransmit the damaged data segments. Therefore, in the case of higher BER in the network, the segment size has to be chosen to be smaller. On the other hand, larger data segments can be used in networks with lower BER. For example, in order to operate at a BER of 10^{-3} a segment size of a few hundred bits should be used; e.g. about 240 bits (30 bytes).

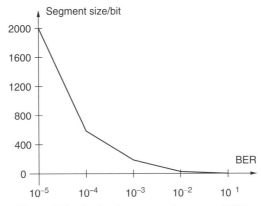

Figure 5.5 Optimal segment size versus BER

The size of data segment is usually chosen to ensure an efficient network operation under the worst acceptable disturbance conditions. However, the BER in a network changes dynamically, depending on several factors, such as number of active stations in the network, activity of noise sources in the network environment, and so on. Thus, the size of the data segments, calculated for the worst case is not optimal any more. Therefore, realization of data segments with variable size, which depends on the current BER in the network, seems to be a reasonable solutions. However, this approach causes a higher complexity for realization of such communications systems.

5.2.2 FDMA

5.2.2.1 Basic FDMA

The next option for the division of the network resources into the accessible sections is to allocate different portions of the available frequency spectrum to different subscribers. This access method is called *Frequency Division Multiple Access*(FDMA). Similar to the orthogonality condition from Eq. (5.1), the orthogonality between different users can also be defined in the frequency range [DaviBe96]:

$$\int_{-\infty}^{\infty} X_i(f) X_j(f) \, df = \begin{cases} 1 & i = j \\ 0 & \text{else} \end{cases} \quad (5.3)$$

FDMA provides a number of transmission channels, representing the accessible sections of network resources, spread in a frequency range (Fig. 5.6). Each transmission channel uses an extra frequency band, within entire frequency spectrum of a transmission medium, that can be allocated to particular users and services. The data rate of a transmission channel depends on the width of the frequency band allocated to the channel. Principally, the transmission channels with both fixed and variable data rates, such as the case in TDMA,

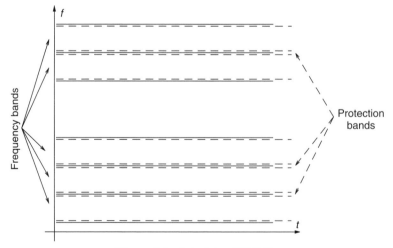

Figure 5.6 Principle of FDMA

can also be realized in an FDMA system by a dynamic frequency allocation to particular transmission channels. To ensure the orthogonality between individual transmission channels, a protection interval in frequency domain has to be provided between FDMA frequency bands.

A big advantage of the FDMA scheme over TDMA is the robustness against narrowband disturbances [MoenBl01] and frequency-selective impulses. In this case, the disturbances can be easily avoided by reallocation of the existing connections from the frequencies affected by the disturbances to the available part of the frequency spectrum. The same principle can be applied for avoidance of the critical frequencies, which are forbidden for PLC because of EMC problems (Sec. 3.3).

FDMA scheme can be implemented in different transmission systems, such as spread-spectrum and OFDM-based transmission systems, which are considered as suitable for realization of broadband PLC systems (Sec. 4.2). In an SS/FDMA system (combination of spread-spectrum and FDMA), the transmission is organized within the frequency bands, provided by the FDMA. On the other hand, because of the specific division of the frequency spectrum in multiple subcarriers, the application of FDMA in OFDM-based transmission systems leads to an OFDMA (OFDM Access) scheme [NeePr00, Lind99, WongCh99], which is also called clustered OFDM [LiSo01]. Because of the robustness of FDMA-based schemes against narrowband disturbances, OFDMA is considered as a suitable solution for the organization of multiple access in PLC access networks.

5.2.2.2 OFDM Access

According to the OFDMA scheme, the subcarriers with relatively low data rates are grouped to build up the transmission channels with higher data rates providing a similar FDMA system [NeePr00, KoffRo02]. However, the protection frequency bands, which are necessary in FDMA to separate different transmission channels (Fig. 5.6), are avoided in an OFDMA system thanks to the provided orthogonality between the subcarriers, as described in Sec. 4.2.1. Each transmission channel (CH) consists of a number of subcarriers (SC), as is presented in Fig. 5.7. The subcarriers of a transmission channel can be chosen to be adjacent to each other, or to be spread out in the available frequency spectrum.

The transmission channels represent the accessible sections of the network resources that are established by the OFDMA scheme. So, the task of the MAC protocol is to manage the channel reallocation between a number of subscribers and different telecommunications services. The transmission channels can be organized so as to have constant or variable data rates, which can be ensured by the association of variable numbers of subcarriers building a transmission channel. The subcarriers can be managed in the following three ways:

(a) A group of subcarriers (SC), all with a fixed data rate, form a transmission channel (CH) with a constant data rate.
(b) A group of subcarriers with variable data rates (caused by bit loading, Sec. 4.2.1) form a channel. Accordingly, the channels also have variable data rates.
(c) The subcarriers are grouped according to the available data rates per subcarrier, in order to build up the transmission channels with a certain data rate. The subcarrier data rates are variable, but the channel data rate remains constant.

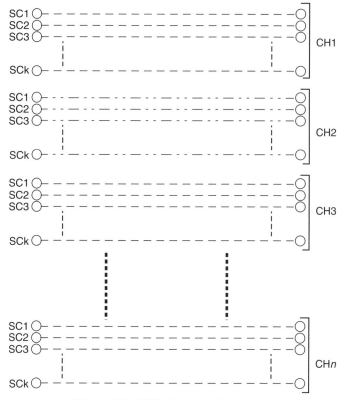

Figure 5.7 OFDMA channel structure

In case A, the transmission channels have the same transmission capacity and always include the same subcarriers (Fig. 5.7). If one or more subcarriers are not available (e.g. they are defective) the transmission channel cannot be used, although other subcarriers are still available. In case B, the subcarriers of a transmission channel change their data rates according to the network and disturbance conditions (bit loading), and with it change the channel data rate, too. In case C, all available subcarriers are summarized into a number of channels with a certain (fixed or variable) transmission capacity. That means, a number of subcarriers are grouped according to their available capacity to form a transmission channel with a desired capacity. In this case, the transmission channels do not always include the same subcarriers.

5.2.2.3 OFDMA/TDMA

As is mentioned above, the slotted nature of OFDM-based transmission systems leads to a logical division of the network resources in the time domain (TDMA). An OFDMA system can also be extended to include the TDMA component, which leads to a combined OFDMA/TDMA scheme (Fig. 5.8). In this case, the transmission channels, which are divided in a frequency range, are also divided into time slots with a fixed or variable duration. Accordingly, each time slot carries a data segment with a fixed or variable

PLC MAC Layer

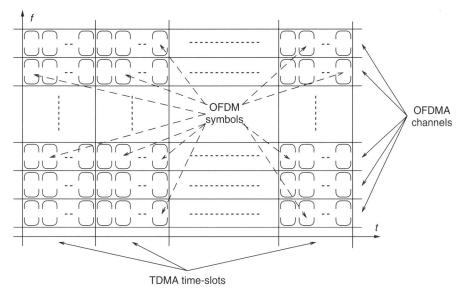

Figure 5.8 OFDMA/TDMA scheme

size. The data segments present the smallest accessible portions of the network resources provided by the OFDMA/TDMA scheme, which are managed by a MAC protocol. Thus, in the case of OFDMA/TDMA, the MAC protocol controls access to both transmission channels and time slots.

Each transmission channel consists of a number of subcarriers, which can be grouped in different ways, as is provided by the OFDMA scheme (Fig. 5.7). Accordingly, a transmission channel can include a variable number of subcarriers or a fixed number of subcarriers with variable data rates (bit loading), causing variable data rates of the transmission channel as well. On the other hand, a time slot carrying a data segment consists of a number of OFDM symbols with a certain duration and payload capacity, as is described above for an OFDM/TDMA system. In any case, the number of the OFDM symbols per time slot and per channel, which corresponds to a data segment, has to be an integer.

5.2.3 CDMA

The CDMA (Code Division Multiple Access) method provides different codes to divide the network resources into the accessible sections. The data from different users is distinguished by the specific code sequences and can be transferred over a same transmission medium, by using a same frequency band, without interferences between them. The CDMA scheme is based on the spread-spectrum principle, recently called *Code Division Multiplex* (CDM), and is also denoted as Spread-Spectrum Multiple Access (SSMA). In Sec. 4.2.2, we presented the spread-spectrum technique from the transmission point of view without consideration of the multiple access capabilities of the CDMA scheme. In the description below, we discuss possibilities to use the features of the spread-spectrum technique for realization of various CDMA systems.

5.2.3.1 Principle

CDMA can be realized by application of several coding methods (see e.g. [Pras98]). The most considered methods in recent telecommunications systems, such as wireless networks, are [DaviBe96, Walke99]

- DS-CDMA – Direct Sequencing CDMA – based on Direct Sequence Spread Spectrum (DSSS) method, where each user's data signals are multiplied by a specific binary sequence, and
- FH-CDMA – Frequency Hopping CDMA – based on Frequency Hopping Spread Spectrum (FHSS) method, where the transmission is spread over different frequency bands, which are used sequentially.

In a DS-CDMA system, all subscribers of a network use the entire available frequency spectrum of a transmission medium. To be able to distinguish between different subscribers, data signals from different network users are multiplied by different code sequences, which are chosen to be unique for every individual user or connection (Fig. 5.9). At the receiver side, the arriving signal is again multiplied by the uniquely specified code sequence. The result of the multiplication is the originally sent data signal, which is extracted between all other data signals, multiplied by different code sequences.

Thus, data signal $S_i(t)$, generated by user i, is multiplied by its corresponding code sequence $C_i(i)$ building a coded signal $S_i(t)C_i(t)$, which is transmitted over a medium (e.g. wireless or PLC channel). A receiving user listens to the transmission medium and can receive coded signals generated by all network users, so-called "signal mix" $S_1(t)C_1(t)$ to $S_n(t)C_n(t)$, originated by application of their own codes. However, to receive and decode the original data signal $S_i(t)$, it is necessary to multiply the signal mix by the unique code sequence $C_i(t)$, which is only known or currently applied by the receiving user.

To explain how it is possible to distinguish between signals from different users in a CDMA system, we present an example by considering two signals $S_a(t)$, with a bit sequence $\{1, 0, 1, 1\}$ and $S_b(t)$, with $\{0, 1, 1, 0\}$, generated by two users A and B (Fig. 5.10). Both users code the bit sequence with their own code sequence $C_a(t)$, with $\{1, 0, 1, 0\}$, and $C_b(t)$, with $\{1, 0, 0, 1\}$, respectively. Both code sequences are transmitted with four times higher data rates than the original user signals.

After the multiplication of bit and code sequences, users A and B deliver their signal products $S_a(t)C_a(t)$ and $S_b(t)C_b(t)$ to a shared transmission medium. Thus, a sum signal $S_a(t)C_a(t) + S_b(t)C_b(t)$ is received by destination users A' and B', which are target users

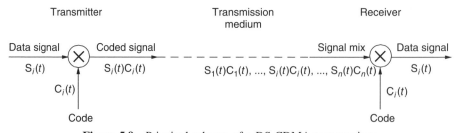

Figure 5.9 Principal scheme of a DS-CDMA transreceiver

PLC MAC Layer

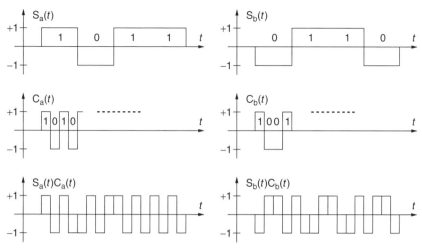

Figure 5.10 CDMA signal generation/coding – example

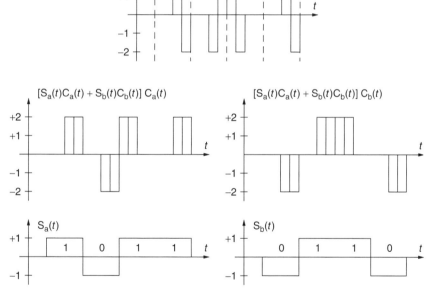

Figure 5.11 CDMA signal decoding – example

for both signals $S_a(t)$ and $S_b(t)$, respectively (Fig. 5.11). To extract the original signals from users A and B at the right receiver, target users A' and B' have to multiply the sum signal by code sequences $C_a(t)$ and $C_b(t)$, which are also used at the transmitters for signal coding. The result of this multiplication is original bit sequences $S_a(t)$ and $S_b(t)$ received by A' and B' respectively.

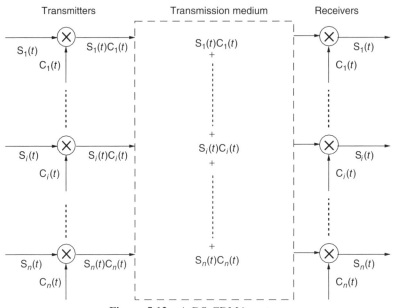

Figure 5.12 A DS-CDMA system

The same principle of dividing information signals of various network users can be applied if a larger number of subscribers use a same shared transmission medium. In this case, a code sequence has to be defined for every connection in the network ($C_1(t), \ldots, C_i(t), \ldots, C_n(t)$), as presented in Fig. 5.12. Both transmitting and receiving participant of a connection have to use the same code sequence. If we consider communications network with a centralized structure, such as PLC access networks (Sec. 3.1), a central unit (e.g. base station) uses a number of code sequences to receive signals from different network users. The application of different codes ensures realization of a transmission channel within a CDMA system. So, the transmission channels are determined by applied code sequences providing the accessible portions of the network resources, such as the time slots in TDMA and frequency bands in FDMA schemes.

As is mentioned above, a DS-CDMA system occupies the entire frequency band that is used for the transmission over a medium. On the other hand, FH-CDMA systems use only a small part of the frequency band, but the location of this part differs in time [Pras98]. During a time interval (Fig. 5.13), the carrier frequency remains constant, but in every time interval, it hops to another frequency (Sec. 4.2.2). The hopping pattern is determined by a code signal, similar as in a DS-CDMA system. Thus, the transmission channels in an FH-CDMA system are defined by the specific code as well. So, during a data transmission, a subscriber uses different frequency bands. The change of the frequency bands in the time is specified by the code sequence, allocated to the subscriber. In a special case, if the codes allocated for the individual users always point to the same frequency band, the same users always transmit over the same frequency bands, which leads to a classical FDMA system.

A further variant of CDMA schemes is TH-CDMA (Time Hopping CDMA), where the data signal is transmitted during so-called "rapid time-bursts" at time interval determined by a specific code sequence (Fig. 5.14). In a TH-CDMA system, the entire frequency

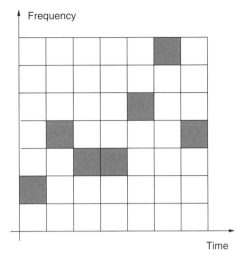

Figure 5.13 FH-CDMA – time/frequency diagram

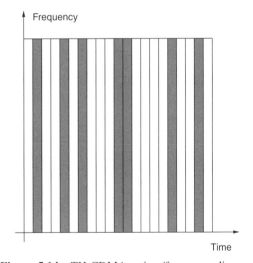

Figure 5.14 TH-CDMA – time/frequency diagram

spectrum is used, such as in a DS-CDMA. However, the exact time slots to be used for a particular transmission are determined by a code sequence, for example, allocated to a network user. If there is a synchronization among code sequences that one user transmits only during a particular time slot, TH-CDMA becomes a TDMA system.

The variants of CDMA presented above can be combined to build up so-called "hybrid CDMA solutions". The hybrid schemes, such as DS/FH, DS/TH, FH/TH and DS/FH/TH, can be applied to join the advantages of different CDMA variants. Furthermore, the CDMA techniques can also be combined with other multiple access schemes; for example, building a CDMA/TDMA [ChlaFa97] or a CDMA/FDMA scheme [SchnBr99]. In a CDMA/TDMA scheme, the accessible sections of the transmission resources are provided by both division in the time domain (by time slots) and division in the code domain, by allocation of code sequences. Thus, a user accesses a determined time slot and applies a specific code sequence. In the case of CDMA/FDMA, the accessible sections are defined by a frequency band (FDMA transmission channel) and a specific code sequence.

Spread-spectrum (SS) can also be combined with multi-carrier modulation (MCM) schemes, such as OFDM, building so-called "multi-carrier spread-spectrum systems" (MCSS)[HaraPr97, FazelPr99, Pras98, Lind99]. MCSS improves the network performances, stabilizing BER and increasing robustness against burst errors. Therefore, MCSS schemes are also considered for the application in PLC [TachNa02].

Multi-carrier spread-spectrum systems can be realized by a combination of frequency domain spreading and MCM, as well as by a combination of time domain spreading and MCM. Accordingly, there are the following basic concepts for realization of multi-carrier multiple access schemes:

- MC-CDMA – Multi-carrier CDMA, where a spread data stream is modulated on the parallel subcarriers so that the chips of a spread data symbol are transmitted in parallel on each subcarrier using the entire frequency spectrum, such as in DS-CDMA (different to pure OFDM system, where only one symbol is transmitted at the same time), and
- MC-DS-CDMA – Multi-carrier DS-CDMA and MT-CDMA – Multi-tone CDMA, where the data is first converted into parallel data stream and after that, direct-sequence spreading is applied to each subcarrier.

A common feature of all these multi-carrier access schemes is that separation of signals from different users is performed in the code domain as well.

5.2.3.2 Orthogonality

As is mentioned above, the orthogonality between transmission channels in TDMA and FDMA schemes has to be provided in time (Eq. (5.1)) and frequency (Eq. (5.3)) domain, respectively. In a CDMA system, transmission channels are defined by used code sequences and the orthogonality between the transmission channels is provided by orthogonality of applied codes. The choice of the type of code sequence is important for the following two reasons [Pras98]:

- Because of multipath propagation effect, that are expected in various communications systems (e.g. PLC and wireless transmission environments), each code sequence has to distinguish from a time-shifted version of itself.
- To ensure multiple access capability of a CDMA communications system, each code sequence, from a code set used in a network, has to distinguish from other codes from the set.

The distinction between two signals or code sequences is measured by their correlation function. Thus, two real-valued signals x and y are orthogonal if their crosscorrelation $R_{xy}(0)$ in a time interval T is zero [Yang98]:

$$R_{xy}(0) = \int_0^T x(t) y(t) \, dt \qquad (5.4)$$

If $x = y$, which means $R_{xy} = R_{xx}$, the Eq. (5.4) represents autocorrelation function of x. In discrete time, the two sequences are orthogonal if their cross-product $R_{xy}(0)$ is zero:

$$R_{xy}(0) = x^T y = \sum_{i=1}^{N} x_i y_i \qquad (5.5)$$

where $x^T = [x_1 x_2 \ldots x_I]$ and $y^T = [y_1 y_2 \ldots y_I]$, representing sequences x and y, and N is code order, which is number of sequence members belonging to a code.

For example, the following two sequences $x^T = [-1 -1 1 1]$ and $y^T = [-1 1 1 -1]$ are orthogonal because their crosscorrelation is zero:

$$R_{xy}(0) = x^T y = (-1)(-1) + (-1)(1) + (1)(1) + (1)(-1) = 0$$

The properties of an orthogonal code set to be used in a CDMA scheme can be summarized as follows [Yang98]:

- The crosscorrelation should be zero, as presented above for codes x and y, or very small.
- Each code sequence has to have an equal number of 1s and -1s, or their number differs by at most 1, which gives a particular code the pseudorandom nature.
- The scaled dot product of each code should be 1.

The dot product of the code x (autocorrelation) is

$$R_{xx}(0) = x^T x = \sum_{i=1}^{N} x_i x_i \qquad (5.6)$$

To get the scaled dot product for the code x, the product from Eq. (5.6) has to be divided by the code order. So, for codes x and y, the scaled dot product is calculated as

$$(x^T x)/N = (x^T x)/4 = (-1)(-1) + (-1)(-1) + (1)(1) + (1)(1) = 4/4 = 1$$
$$(y^T y)/N = (y^T y)/4 = (-1)(-1) + (1)(1) + (1)(1) + (-1)(-1) = 4/4 = 1$$

In a transmission system where multipath signal propagation problem exists, such as PLC networks, it is possible that so-called "partial correlation" between orthogonal code sequences occurs. This problem comes especially in networks with nonsynchronized transmitters. However, even if the transmitters are synchronized, there are varying propagation delays of signals from different transmitters, as well as a same transmitter caused by the multipath signal propagation.

If we consider two succeeding code sequences of the codes x and y, defined above, it can be recognized that they are orthogonal (in accordance with Eq. (5.5)) if they are perfectly aligned [Yang98]:

$$x_i : -1 \ -1 \ +1 \ +1 \ -1 \ -1 \ +1 \ +1$$
$$y_i : -1 \ +1 \ +1 \ -1 \ -1 \ +1 \ +1 \ -1.$$

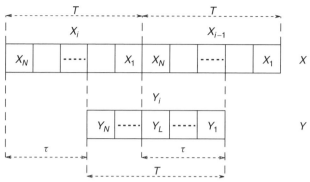

Figure 5.15 Shifted code sequences

However, if the code sequence y delays for any reason for one chip duration (duration of one sequence member), these two codes are no longer orthogonal:

$$x_i : -1\ -1\ +1\ +1\ -1\ -1\ +1\ +1$$
$$y_{i-1} : +1\ +1\ -1\ -1\ +1\ +1\ -1\ -1.$$

To consider a general case, we observe two code sequences x and y, which are shifted for a certain delay τ (Fig. 5.15). The following two partial correlation functions can be defined [Pras98]:

$$R_{xy}(\tau) = \int_0^\tau x(t)y(t-\tau)\,dt \tag{5.7}$$

$$R_{xy}(\tau) = \int_\tau^T x(t)y(t-\tau)\,dt = \int_\tau^{NT_c} x(t)y(t-\tau)\,dt \tag{5.8}$$

Code period can be expressed as $T = NT_c$, where T_c is duration of a code chip. As is also mentioned above, if $x = y$ then Eqs. (5.7) and (5.8) represent the partial autocorrelation functions.

If we assume that τ is a multiple of the chip duration, implying $\tau = LT_c$, the partial correlation functions (Eqs. (5.7) and (5.8)) can be written as

$$R_{xy}(L) = \sum_{i=1}^{L} x_i y_{i-L}, \tag{5.9}$$

and

$$R_{xy}(L) = \sum_{i=L+1}^{NT_c} x_i y_{i-L} \tag{5.10}$$

respectively.

It can be concluded that the simple orthogonality between two aligned code sequences is not enough to ensure always the distinction between the codes and accordingly coded data patterns. Both partial correlation functions have to be zero as well or, at least, very small, for any value of the delay τ, which is expected in a communications network [Yang98]. Furthermore, the same can be concluded for the partial autocorrelation functions, which is necessary to reduce the effect of the multipath propagation and following interference between time-shifted versions of a same coded sequence.

5.2.3.3 Generation of Code Sequences

A Pseudo-Noise Sequence (PNS) acts as a noise-like, but deterministic, carrier signal used for bandwidth spreading of the information signal energy. The selection of a suitable code is of a primordial importance, because the type and the length of the code determines the performances of the system. The PNS code is a pseudo-noise or pseudorandom sequence of ones and zeros, but is not real random sequence because it is periodic and because identical sequences can be generated if the initial conditions or value of the generator are known. The basic characteristic of a PNS is that its autocorrelation has properties similar to those of the white noise, whose energy is constant over the entire occupied frequency spectrum. The autocorrelation $R_{a,\text{WGN}}$ of a White Gaussian Noise (WGN) and its Fourier transform, representing the signal energy over the spectrum, is illustrated in Fig. 5.16. The generated PNSs have to near these properties.

For PNS, the autocorrelation has a large peaked maximum, Fig. 5.17, only for perfect synchronization of two identical sequences, like white noise. The synchronization of the receiver is based on this property. The frequency spectrum of the PN sequence has spectral lines that become closer to each other with increasing sequence length N; this is because of the periodicity of the PNS. Each line is further smeared by data scrambling, which spreads each spectral line and further fills in between the lines to make the spectrum more nearly continuous, [Meel99b]. The DC component is determined by the zero-one balance of the PNS.

The crosscorrelation $R_{xy}(\tau)$ describes the interference between two different codes x and y, by measuring agreement between them. When the crosscorrelation is zero for all τ, the user codes are called *orthogonal* and therefore there is no interference between the users after the de-spreading and the privacy of the communication for the users is kept. However, in practice, the codes are not perfectly orthogonal. Hence, the crosscorrelation between user codes introduces performance degradation, by increased noise power after de-spreading, which limits the maximum number of simultaneous users.

In the practice, a wide range of PNS generator classes are implemented. In the following, the mostly encountered ones are described; [Meel99b]:

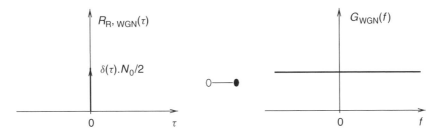

Figure 5.16 Autocorrelation of the White Gaussian Noise

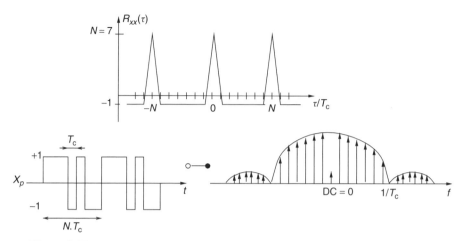

Figure 5.17 Autocorrelation and the frequency occupation of a periodic sequence

m-Sequence Codes

A Simple Shift Register Generator (SSRG) has all the feedback signals returned to a single input of a shift register (a delay line), as presented in Fig. 5.18. The SSRG is linear if the feedback function can be expressed as a modulo-2 sum, through X-OR ports. In this case, this generator is also called Linear Feedback Shift Register (LFSR).

The feedback function $f(x_1, x_2, \ldots, x_n)$ is a modulo-2 sum of the contents x_i of the shift register cells with c_i being the feedback connection coefficients, where $c_i = 1 =$ connect and $c_i = 0 =$ open.

An SSRG generator with L flip-flops produces sequences that depend on register length L, feedback tap connections and initial conditions. When the period (length) of the sequence is exactly $N = 2^L - 1$, the PN sequence is called a *maximum-length sequence* or simply an *m-sequence*. If an L-stage SSRG has feedback taps on stages L, k, m and has sequence "$\ldots, a_i, a_{i+1}, a_{i+2}, \ldots$", then the "reverse SSRG" has feedback taps on L, $L - k, L - m$ and sequence "$\ldots, a_{i+2}, a_{i+1}, a_i, \ldots$", see Fig. 5.19.

For the balance of an m-sequence, there is one more "ones" than "zeros" in a full period of the sequence. Since all states but the "all-zero" state are reached in an m-sequence, there must be 2^{L-1} "ones" and $2^{L-1} - 1$ "zeros". For every m-sequence period, half the runs (of all 1's or all 0's) have length 1, one-fourth have length 2, one-eighth have length 3, and so on. For each of the runs, there are equally many runs of 1's and 0's.

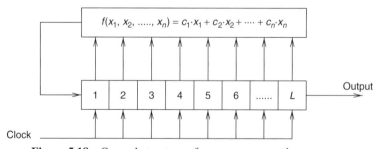

Figure 5.18 General structure of a *m*-sequence codes generator

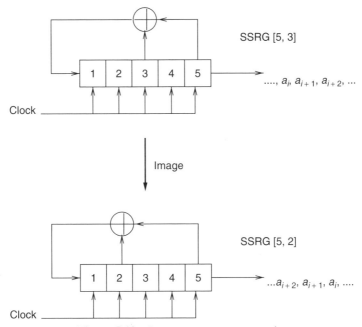

Figure 5.19 Reverse sequence generation

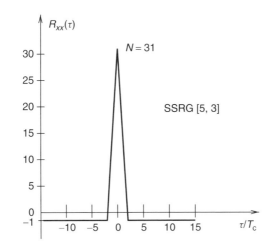

Figure 5.20 Autocorrelation of the m-sequence codes

The autocorrelation function of the m-sequence is "−1" for all values of the chip phase shift τ, except for the [−1, +1] chip phase shift area, in which correlation varies linearly from the "−1" value to $2^{L-1} = N$, which is the sequence length, as illustrated in Fig. 5.20. The autocorrelation peak increases with increasing length N of the m-sequence and approximates the autocorrelation function of white noise. This is the unique advantage of the m-sequence toward all other PNS codes generators. Unfortunately, its crosscorrelation is not as good as its autocorrelation. Therefore, when a large number of transmitters using different codes share a frequency band, the code sequences must be carefully chosen to avoid interference between users.

Gold Codes

In spite of its best autocorrelation properties, the m-sequence generator cannot be optimally used in a CDMA environment, because a multiuser system needs a set of codes with the same length and with good crosscorrelation characteristics. Gold code sequence generator is very useful in such environment because a large number of codes, with the same length and with controlled crosscorrelation, can be generated. Furthermore, this realization is possible with only one pair of feedback tap sets.

Gold codes can be generated by the modulo-2 adding, through an exclusive OR, of two maximum-length sequences with the same length N, with $N = 2^r - 1$, where r odd or $r = 2 \bmod 4$. The code sequences are added chip by chip by synchronous clocking, as illustrated in Fig. 5.21 for the general structure and in Fig. 5.22 for an example. Because the m-sequences are of the same length, the two code generators maintain the same phase relationship and the generated Gold codes have the same length as their m-sequence basic codes, but are not maximal. Therefore, the Gold sequences autocorrelation function will be worse than that of the m-sequence codes, as shown in the example illustrated in Fig. 5.23. A 2-register Gold code generator of length L can generate $2^L - 1$ sequences plus the two base m-sequences, which gives a total of $2^L + 1$ sequences.

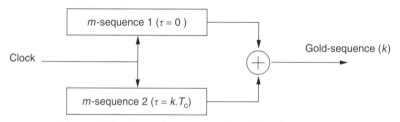

Figure 5.21 General structure of a gold codes generator

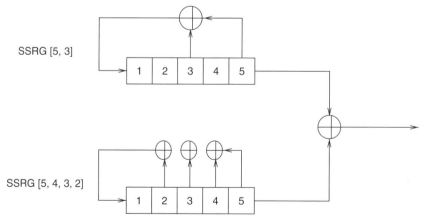

Figure 5.22 Example of gold codes generators

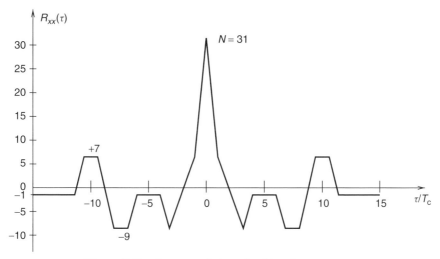

Figure 5.23 Crosscorrelation of gold codes sequences

In addition to their advantage to generate large numbers of codes, the Gold codes may be chosen so that over a set of codes available from a given generator, the autocorrelation and the crosscorrelation between the codes is uniform and bounded. If specially selected m-sequences, called *preferred pair PN m-sequences*, are used, the generated Gold codes have a three-valued crosscorrelation. In this case, the autocorrelation can be expressed by [FleuKo02]:

$$R_{xx}(\tau) \begin{cases} = N, & \text{if } \tau = 0 \\ \in \{-t(r), -1, t(r) - 2\} & \text{otherwise} \end{cases} \quad (5.11)$$

and the crosscorrelation

$$R_{xy}(\tau) \in \{-t(r), -1, t(r) - 2\} \quad (5.12)$$

where

$$t(r) \begin{cases} 1 + 2^{\frac{r+1}{2}}, & \text{for } r \text{ odd} \\ 1 + 2^{\frac{r+2}{2}}, & \text{for } r = 2 \bmod 4 \end{cases} \quad (5.13)$$

and for a large N, the crosscorrelation bound is expressed as

$$\max |R_{xy}(\tau)| = |t(r)| \approx \begin{cases} \sqrt{2} \cdot 2^{\frac{r}{2}} = \sqrt{2} \cdot R_{xx}, & \text{for } r \text{ odd} \\ 2 \cdot 2^{\frac{r}{2}} = 2 \cdot R_{xx}, & \text{for } r = 2 \bmod 4 \end{cases} \quad (5.14)$$

The Gold code generator presented in Fig. 5.22 is realized by $r = 5$ registers, then the maximum-length sequences have length $N = 2^r - 1 = 31$ and the $R_{xx}(\tau = 0) = N$. Furthermore, the number r is an odd number, then the autocorrelation for τ different to zero takes the values from the set $\{-9, -1, +7\}$ according to Eq. (5.11), because $t(r) = 9$ according to Eq. (5.13). This autocorrelation function is presented in Fig. 5.23.

5.2.3.4 Capacity

In TDMA and FDMA systems, network capacity is limited by used frequency spectrum determining the number of the transmission channels in time and frequency domain, respectively. In CDMA systems, theoretically it is possible to realize an infinite number of channels by allocating different code sequences to each channel. However, the network capacity in CDMA systems is also limited according to the used frequency spectrum and the number of transmission channels is limited as well. To analyze capacity in networks with CDMA schemes, we consider the amount of CDMA network capacity by consideration of the amount of interfering users in the available frequency band, presented in [Yang98].

Performance of different digital modulation and transmission schemes depends on so-called "link metric" E_b/N_0, or energy per bit per noise power density. Energy per bit can be defined as average modulating signal power (S) allocated to each bit duration (T), that is $E_b = ST$. If the bit duration is substituted by bit rate R, which is inverse of the bit duration T, the energy per bit is $E_b = S/R$. So, the link metric can be written as

$$\frac{E_b}{N_0} = \frac{S}{RN_0} \tag{5.15}$$

The noise power density is the total noise power divided by the used frequency spectrum - bandwidth $N_0 = N/W$. Substituting it in Eq. (5.15), the link metric is

$$\frac{E_b}{N_0} = \frac{S}{N}\frac{W}{R} = \text{SNR}\frac{W}{R} \tag{5.16}$$

dividing the energy per bit in two factors: signal-to-noise ratio and processing gain of the system (W/R). If we assume that the system possesses perfect power control, which means that received signal power from all network users is the same, SNR of one network user can be written as

$$\text{SNR} = \frac{1}{M-1} \tag{5.17}$$

where M is total number of users in the network. Thus, the interference power in the used frequency band is equal to the sum of powers from individual users, as presented in Fig. 5.24. However, Eq. (5.17) ignores other interference sources, such as thermal noise, influence of neighboring communications systems, and so on.

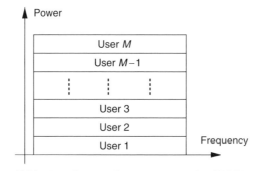

Figure 5.24 Interferences between users of a CDMA system

Substituting Eq. (5.17) into Eq. (5.16), the link metric is

$$\frac{E_b}{N_0} = \frac{1}{(M-1)} \frac{W}{R} \qquad (5.18)$$

Solving Eq. (5.18) for $(M-1)$, it is

$$M - 1 = \frac{(W/R)}{(E_b/N_0)} \qquad (5.19)$$

If $M \gg 1$ the total number of users M in the CDMA network is

$$M = \frac{(W/R)}{(E_b/N_0)} \qquad (5.20)$$

In accordance with Eqs. (5.19) and (5.20), it can be concluded that the number of users simultaneously using network resources is directly proportional to the processing gain of the system (W/R). On the other hand, the lower the required threshold for the energy per bit per noise power density, the higher is the network capacity. So, the maximum number of users in the network is inversely proportional to the required link metric (E_b/N_0).

If we consider communications system with frequency reuse, such as cellular mobile networks and broadband PLC access networks with repeaters (Sec. 2.3.3 and Sec. 3.1), a CDMA-based network cannot be considered as an isolated system, because it is influenced by neighboring network segments or cells. In this case, a network segment is said to be loaded by the neighboring systems, reducing its capacity. Accordingly, Eq. (5.20) is modified to include so-called "loading factor" η, with a value range between 0 and 1 (Eq. (5.21)),

$$M = \frac{(W/R)}{(E_b/N_0)} \left(\frac{1}{1+\eta}\right) = \frac{(W/R)}{(E_b/N_0)} F \qquad (5.21)$$

where F, as the inverse of $(1+\eta)$, is known as frequency reuse factor [Yang98].

On the other hand, the users of a network applying various telecommunications services do not transmit data for the entire duration of their connections with a constant data rate, as is discussed in Sec. 4.4. Even if packet voice service is considered, the speech statistics show that a user in a conversation typically speaks between 40 and 50% of the time. Such transmissions with variable data rates reduce the total interference power in a CDMA system by so-called "voice activity factor" v. This increases the network capacity, as is shown by extension of Eq. (5.21) for the activity factor in Eq. (5.22).

$$M = \frac{(W/R)}{(E_b/N_0)} \left(\frac{1}{1+\eta}\right) \left(\frac{1}{v}\right) \qquad (5.22)$$

In accordance with Eq. (5.21) and Eq. (5.22), it can be concluded that the capacity of a CDMA system also depends on the influences from the network environment (loading) and characteristics of currently transmitted data patterns (from services with variable data rates).

In TDMA and FDMA systems, number of transmission channels, with fixed or variable data rates, is firmly determined by the number of time slots or frequency bands. If there are no free transmission channels in a network, new connections cannot be accepted, causing so-called "blocking". In CDMA systems, the same situation exists if there are no free channels (codes) in the network, causing so-called "hard blocking". However, CDMA systems allow an increase of the number of users so far as the level of interferences is still acceptable. If it is not the case, the interferences negatively affect the QoS in the network and we talk about so-called "soft blocking", which is a particularity of the CDMA systems.

To analyze the soft blocking, we consider a simplified model, based on a soft blocking model presented in [Yang98]. Total interference in a CDMA network can be represented as

$$I_{\text{total}} = ME_b R + N.$$

A soft blocking occurs when the total interference level exceeds the background noise level by a predetermined amount $1/r (I_{\text{total}} = N/r)$. Thus, the soft blocking occurs when

$$I_{\text{total}} \geq ME_b R + N \qquad (5.23)$$

Substituting $N = I_{\text{total}} r$ and $I_{\text{total}} = WI_0$ in Eq. (5.23), where I_0 is interference power density, it results with

$$WI_0 \geq ME_b R + rWI_0 \qquad (5.24)$$

Solving Eq. (5.24) for M, maximum number of users in the system is given by Eq. (5.25).

$$M = \frac{(W/R)}{(E_b/N_0)}(1 - r) \qquad (5.25)$$

It can be concluded that the capacity of a CDMA system is function of a maximum tolerable bit error rate due to Multiple Access Interference (MAI). So, the maximum number of active users in a network has to be defined that level of MAI is just below the maximum tolerable. This depends on the system features, such as number of receivers, degree and type of the code set, and properties of used MAC protocol [JudgTa00].

The transmission channels provided by the CDMA scheme can be with fixed or variable data rates, such is the case in TDMA and FDMA schemes. Realization of channels with the variable data rates can be done by adapting the spreading code, allocated to the transmission channel, or by a change of the (frequency) bandwidth, occupied by the channel. Another way to achieve the variable data rates is transmission of a data stream belonging to a logical transmission channel by using multiple codes allocated to a user. However, the last solution is not efficient and increases complexity of CDMA receivers [Walke99].

5.2.4 Logical Channel Model

As is presented above, all three multiple access schemes provide so-called "accessible sections" of the network resources in time domain (TDMA), by an amount of time slots within repeating time frame, in frequency domain (FDMA), by a number of allocated frequency bands, and in code domain (CDMA), by allocation of orthogonal code sequences for different signals that are transmitted at the same time using a same frequency bandwidth. Independent of the applied multiple access scheme, a communications system provides so-called "transmission channels" (accessible sections) that are used by multiple

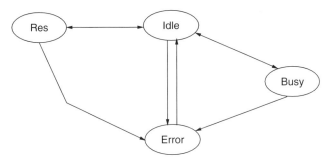

Figure 5.25 Simple channel state diagram

subscribers applying various telecommunications services. Accordingly, it is possible to set up a general channel model representing the transmission resources of a communications network using any multiple access scheme (Fig. 5.25).

Generally, a transmission channel is in busy state if it is used for any kind of transmission. It can also be in an idle state (free), in an error state (disturbed), or reserved (Res). Idle channels can be allocated to new connections in the network. If the channels are disturbed, they are in the error state. After the disturbance disappears, the channels are again idle. A special pool of the transmission channels can be in a reserved state. These channels are reserved for the substitution of currently used channels, which are affected by the disturbances ensuring continuation of existing connections, or to ensure an immediate acceptance of connections with a higher priority.

Transitions from reserved, idle and busy states to the error state (Fig. 5.25), as well as from the error state to the idle channel state are caused by disturbances, produced by various types of noise. The disturbances and the resulting state transitions can be modeled by an on–off model, as presented in Sec. 3.4.4. However, the transmission channels provided by different multiple access schemes react differently to the disturbances in accordance with their duration, frequency occupancy and power. So, a frequency-selective disturbance impulse can affect only a number of transmission channels in an FDMA system, whereas all time-slots of a TDMA system are in the error state for the entire impulse duration.

On the other hand, the task of the MAC layer and its protocols is to control the transitions between possible channel states, besides the error state. This is carried out by MAC protocols and traffic control mechanisms in accordance with the current traffic and disturbance situation in the network.

5.3 Resource-sharing Strategies

The task of the resource-sharing strategies – MAC protocols – is to organize the access of multiple subscribers using the same, shared network resources, which is carried out by managing the accessible sections of the network transmission resources provided by a multiple access scheme (Sec. 5.2). The organization of the transmission in the downlink direction seems to be easy because it is fully controlled by the base station (Fig. 5.26). In this direction, the base station transmits data to one or multiple network stations, or it broadcasts information to all network stations. In any case, there are only data packets

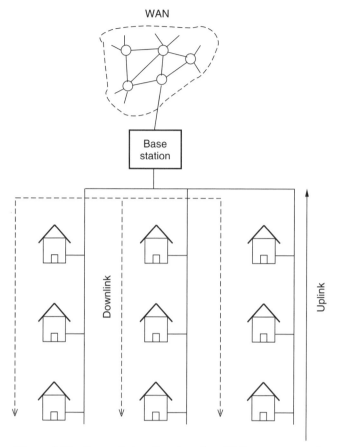

Figure 5.26 Transmission directions in a PLC access network

from the base station on the medium and no synchronization between transmissions of different network stations is necessary in the downlink.

On the other hand, multiple network stations have to compete for medium access in the uplink. The network stations operate independently and each station can have data to transmit at any time. Therefore, the transmission in the uplink has to be organized by a MAC protocol to ensure a fair network usage for all network stations and to prevent collisions between data packets transmitted from different network stations.

The point of interest in this section is the investigation of MAC protocols to be applied to the PLC uplink according to the requirements of PLC networks, discussed in Sec. 5.1.3. For this purpose, we analyze various protocol variants. Beginning from simple ALOHA protocols, we present the particularities of random access principle and describe various extensions of the random protocols, which can improve network performance. Furthermore, arbitration protocols, such as polling, token- passing and reservation, are analyzed for their application in PLC as well. Recent broadband PLC access networks apply variants of Carrier Sense Multiple Access (CSMA) protocol and reservation MAC protocols. Therefore, we pay attention on performance analysis of the CSMA protocols and describe in detail one of its extended implementation variants, IEEE 802.11 MAC protocol. A comprehensive performance evaluation of the reservation protocols for PLC is separately presented in Chapter 6.

5.3.1 Classification of MAC Protocols

MAC protocols can be divided into two main groups: protocols with a fixed or a dynamic access. The fixed access schemes assign a predetermined fixed capacity to each subscriber for the entire duration of a connection, as is the case in classical telephony. The assigned network capacity is allocated for a subscriber independent of its current need for a certain data rate. Thus, if internet access is used, the allocated network capacity remains unused during viewing phase (Sec. 4.4.2), when no data is transmitted over the network causing so-called "transmission gaps", as shown in Fig. 5.27. On the other hand, the bursty characteristic of a data stream can cause so-called "transmission peaks", when capacity of the allocated channel is not enough to serve the data burst, causing additional delays and decreasing data throughput. For these reasons, the fixed strategies are suitable only for continuous traffic, but not for bursts of data traffic (bursty traffic) [AkyiMc99], typical for different kinds of data transfer that are expected in the access networks, such as broadband PLC networks.

Unlike fixed access methods, dynamic access protocols are adequate for data transmission, and in some cases, it is also possible to ensure realization of QoS guarantees for various telecommunications services. The dynamic protocols are divided into two subgroups; contention and arbitration protocols (Fig. 5.28). In accordance with the contention access principle, the network stations access the transmission medium randomly, which can cause collisions between data units of different network users. Note that a network station does not have knowledge about transmission needs of other stations. So, if two or more stations start to transmit their data packets at the same time, a collision will occur. On the other hand, the arbitration protocols provide a coordination between the network stations, ensuring a dedicated access to the medium. In this case, the network stations access the medium in a determined manner, avoiding the collisions. However, the arbitration procedure takes an additional time, causing longer transmission delays in the network.

Basic protocol solutions, such as ALOHA and CSMA random access methods, as well as token-passing and polling arbitration protocols, can be extended to improve their performance. Thus, the random protocols can be extended by implementation of various mechanisms for collision resolution to reduce number of collisions in the network,

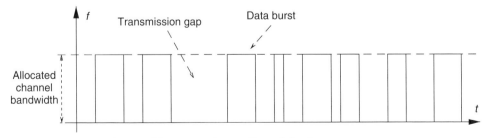

Figure 5.27 Bursty data traffic and fixed access strategy

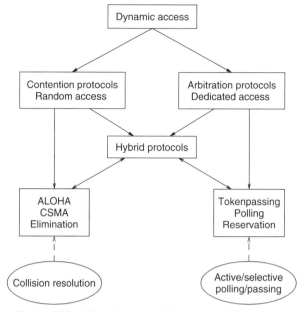

Figure 5.28 Classification of dynamic MAC protocols

whereas the arbitration can be carried out selectively, in accordance with current traffic situation in a network, to reduce the transmission delays. Furthermore, the contention and the arbitration protocols can also be combined to build up so-called "hybrid protocol solutions". The aim of the hybrid protocols is to join advantages of different access methods to improve network performance and to ensure realization of QoS guarantees for various telecommunications services.

5.3.2 Contention Protocols

5.3.2.1 ALOHA Protocols

The pure ALOHA protocol is one of the first introduced access techniques for application in data networks [Chan00, Pras98, RomSi90]. It is characterized by a low realization complexity and a simple operation principle. According to the pure ALOHA protocol, a network station with a packet to transmit simply tries to send it without any coordination with other network stations. Therefore, it is possible that more than one station transmit the packets simultaneously, which causes packet collisions. Thus, the packets generated by two different network stations A and B collide if they are transmitted at the same time (PG – packet generation, Fig. 5.29). In this case, the overlapping data packets from both stations are destroyed. After a collision, the stations try to retransmit the packets after a randomly calculated waiting time. The retransmitted packets can also collide, causing new retransmissions.

To analyze performance of the pure ALOHA protocol, we assume that the packets from different network users are generated in accordance with the Poisson arrival process with

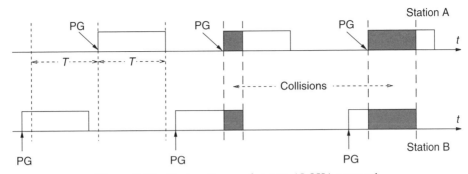

Figure 5.29 Timing diagram for pure ALOHA protocol

a rate g, which is referred as offered load to the channel [RomSi90]. If we consider a packet to be transmitted over the network at a moment t, either as a new or a retransmitted packet, its transmission will be successful if there are no generated packets (Fig. 5.29) from other network stations within the interval $[T - t, T + t]$, where T is duration of the transmitted packet (time needed for packet transmission over considered channel). Since the packet generation is a Poisson process, the probability that no packets are generated within an interval with length $2T$ is

$$P_{suc} = e^{-2gT} \tag{5.26}$$

which represents the probability of the successful packet transmission as well. That means the packets are generated with rate g, but only a fraction of the packets with the probability P_{suc} are successfully transmitted. Accordingly, the rate of successfully transmitted packets is gP_{suc} and each of the packets, carrying useful information, occupies the channel for time T. Using definition that network utilization is the fraction of time that useful information is carried by a transmission channel provided by the network, we can write

$$S = gTP_{suc} \tag{5.27}$$

Note, that the network utilization S (Eq. (5.27)) is very often defined as throughput. However, because of a better understanding of the performance analysis of reservation MAC protocols, presented in Chapter 6, we chose network utilization as a term for network throughput and data throughput as a term describing throughput of individual network stations.

Product $G = gT$ is defined as normalized offered load to the channel [RomSi90]. So, the network utilization can be written as

$$S = GP_{suc} \tag{5.28}$$

If theoretically is possible that $P_{suc} = 1$, which means that every packet is successfully transmitted, the network utilization is equal to the normalized offered load $S = G$. Accordingly, a normalized offered load $G = 1$ corresponds to the maximum network utilization $S = 1$, which means that offered network load (e.g. expressed in bps) of 1 is equal to the maximum data rate in the network.

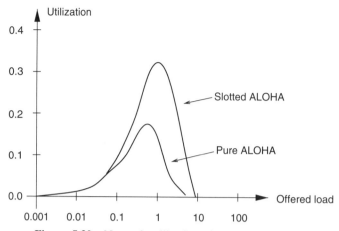

Figure 5.30 Network utilization of ALOHA protocols

Substituting Eq. (5.26) and $G = gT$ in Eq. (5.28), we have finally

$$S = Ge^{-2G} \qquad (5.29)$$

The random nature of the ALOHA protocol causes a very low network utilization (maximum 18%) as is shown in Fig. 5.30. Additionally, ALOHA protocols are characterized by an instable behavior with a resulting performance collapse (network utilization is almost zero) if the network is highly loaded, which makes the realization of QoS guarantees difficult. For these reasons, it can be concluded that the pure ALOHA protocol is not suitable for application in PLC access networks.

Performance of pure ALOHA protocol can be improved by application of so-called "slotted ALOHA protocol", where the transmission channel is divided into time slots, whose size equals the duration of a packet transmission T. The network stations can start transmission of a packet only at the beginning of a time slot (PT – packet transmission, Fig. 5.31). Thus, after generation of a packet (PG) station A has to wait for beginning of next time slot to transmit the packet. Therefore, there is no collision between second packets of stations A and B, as was the case in pure ALOHA protocol (Fig. 5.29). In slotted ALOHA protocol, a collision occurs only if two or more network stations transmit a packet in the same time slot (Fig. 5.31), as is the case with third packets of stations A and B.

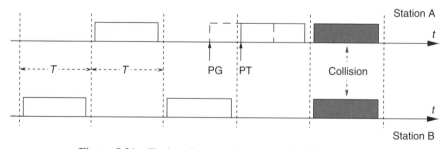

Figure 5.31 Timing diagram for slotted ALOHA protocol

In accordance with the slotted ALOHA, a packet will be successfully transmitted if no other packets are generated within a time slot with the duration T. If we assume that the packet generation is a Poisson process, the probability for a successful packet transmission is.

$$P_{\text{suc}_{S-\text{ALOHA}}} = e^{-gT} \tag{5.30}$$

Substituting Eq. (5.30) in Eq. (5.27), we get network utilization for slotted ALOHA protocol

$$S_{S-\text{ALOHA}} = gTe^{-gT} = Ge^{-G} \tag{5.31}$$

The slotted ALOHA achieves much better network utilization (36%) than the pure ALOHA protocol (Fig. 5.30). The maximum utilization of slotted ALOHA is achieved at $G = 1$, whereas the pure ALOHA achieves its maximum at $G = 1/2$. However, the same instable performance behavior still remains and two basic requirements on the MAC protocol for PLC access networks are not fulfilled by the slotted ALOHA as well (good network utilization, QoS guarantees for various services).

5.3.2.2 Methods for Collision Resolution

By observing network utilization of ALOHA protocols in dependence of generalized offered load to the network, it can be recognized that if the network is highly loaded, the utilization decreases dramatically almost to a zero value. In the highly loaded network, there is a larger number of active network stations, transmitting data packets, or/and a higher number of packets to be transmitted over the network. Therefore, the number of collisions in the network increases significantly, causing a high number of the retransmissions and additionally increasing total number of the packets to be transmitted. In such a situation, only a few number of packets can be transmitted successfully, which results with a minor network utilization. Furthermore, a high collision probability in the network causes frequent packet retransmissions, and accordingly longer transmission delays for the successful data packets.

To improve performance of ALOHA protocols as well as other protocols based on random access, it is necessary to reduce the collision probability in the network. A possibility to solve this problem is application of so-called "mechanisms for collision resolution" (also called Collision Resolution Protocols – CRP) as an additional feature of random access protocols. There is a large number of different proposals for the collision resolution mechanisms, which can be found in the literature and in actual publications regarding MAC layer and protocols in modern telecommunications networks. To present some general solutions and ideas for the collision resolution mechanisms, we divide the resolution methods in following three groups:

- Dynamic backoff mechanisms, ensuring a change of time interval used for calculation of a random moment for transmission of a new or a retransmitted packet, according to the actual number of collisions in the network,
- Calculation of an optimal retransmission probability, in accordance with current situation in the network; collision probability, number of active stations, network load, and
- Collision resolving, carried out by mechanisms for a fast resolution of collisions in the network after they occur.

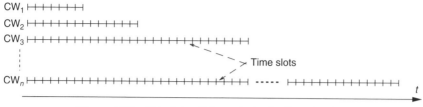

Figure 5.32 Principle of dynamic backoff mechanism

Dynamic Backoff Mechanism
The application of dynamic backoff mechanisms is a very simple method for the reduction of the collision probability, which is also applied in IEEE 802.3 Ethernet-LAN standard. The principle of the dynamic backoff mechanism can be explained as follows: after a first unsuccessful transmission (collision), the affected network station sets a contention window (CW) on a default value (e.g. CW 1, Fig. 5.32). The retransmission of the collided packet will be carried out in a randomly calculated moment within the CW. If the retransmission is also unsuccessful, the CW is increased and a time for the next retransmission is calculated within the new CW. This procedure is repeated until the packet is successfully transmitted. The transmission of a new packet starts again with the default CW, or with the last used CW, which depends on the specific variant of an applied dynamic backoff mechanism.

The increase of the contention window reduces the collision probability, because the probability that two or more network stations transmit at the same time slot decreases with the increase of the CW. Even in the case that a higher number of stations are currently retransmitting the packets (backlogged), the contention window can be increased so far that the collision probability becomes very small.

The increase of the CW can be carried out according to the exponential backoff mechanism, for example, as described in [Walke99], or any other algorithm. In accordance with the exponential backoff mechanism, the access to the channel is controlled by an access probability for each network station. The access probability is determined as $p = 2^{-i}$, where i is number of collisions for a data packet. Thus, for each retransmission attempt, the access probability is equally distributed within a time slot interval $[1, 2^i]$, representing a contention window CW_i (Fig. 5.32). In this way, the contention window is extended for every increase of variable i, representing a new packet collision.

The application of the dynamic backoff mechanism stabilizes random protocols and avoids the performance collapse in highly loaded networks. However, the maximum network utilization is not significantly increased. On the other hand, the increase of the contention window causes longer transmission delays. Therefore, there are some limits for a maximum CW regarding the transmission delays. Accordingly, realization of QoS guarantees for time-sensitive telecommunications services seems to be difficult as well.

Calculation of Optimal Retransmission Probability
A further possibility for the collision resolution is the calculation of the transmission/retransmission probability for the packets in accordance with the current load situation in the network. It can be carried out by an estimation of the backlog – number of collided packets – and the calculation of an adequate transmission probability to avoid the collisions. Several stabilization algorithms are described in [Walke99]; for example, the

Pseudo-Bayesian Algorithm, based on a method that establishes the estimated value of backlog in order to stabilize the slotted ALOHA protocols, also considered in [FrigLe01, FrigLe01a, ZhuCo01], and Minimum Mean-Squared Error algorithm estimating the number of collided stations.

To stabilize a slotted ALOHA protocol, the task of the Pseudo-Bayesian Algorithm is to estimate number of backlog or network station n attempting to transmit new packets or retransmit collided packets. Then each packet will be transmitted with the probability $P = \min\{1, 1/n\}$.

The minimum operation established an upper limit for the transmission probability and causes the access rate $G = np$ to become 1 [Walke99].

A common problem of different stabilization algorithms is the calculation of an optimal retransmission probability because of dynamic load conditions in the network. Thus, number of backlog n depends on the arrival rate, as well as number of active stations in the network. If the calculation of an optimal transmission probability is carried out by a central instance, for example, base station, there is an additional overhead information to be exchanged between network stations, causing an extra signaling network load. Furthermore, in communications networks operating under unfavorable disturbance conditions, such as PLC, signaling messages can be frequently destroyed, which can also influence transmission of information, necessary for calculation of the optimal retransmission probability, between base and network stations. Finally, despite complexity of such algorithms, the result of their application is only a performance stabilization, such as in the case of the dynamic backoff mechanism, described above.

To avoid the signaling exchange between base and network stations, in Rivest's Pseudo-Bayesian algorithm, every node estimates the number of backlog n, and accordingly adjusts its transmission probability $P = 1/n$ [ZhuCo01]. An update of the value n is carried out in accordance with the following rules:

- If a transmission was successful or a contention slot was idle: $n = n - 1$, if $n > 1$, or
- After a collision: $n = n + (e - 2)^{-1}$.

So, it can be concluded that this variant of pseudo-Bayesian algorithm operates similar to the dynamic backoff mechanism, described above.

Collision Resolving

Both collision resolution mechanisms, presented above, are based on an adaptation of the transmission probability, explicitly calculated like in Pseudo-Bayesian algorithm or by a dynamical change of the contention window in the dynamic backoff mechanism, according to the current load situation or the collision probability in a network. Thus, by application of these mechanisms it is tried to avoid the collisions in next contention intervals. A third method for collision resolution can be defined as a procedure for collision resolving, which is carried out after a collision have been occurred.

As an example of collision-resolving mechanisms, we consider a splitting algorithm, which divides the backlogged network stations into subsets, so far that all collisions are resolved [Walke99, RomSi90]. After a collision, all stations involved are divided into two subsets, according to a binary splitting algorithm. Each of the subsets can contain a number of collided stations or it can contain no collided stations. The stations of a subset get allocated an extra portion of the network capacity to retransmit the collided packets.

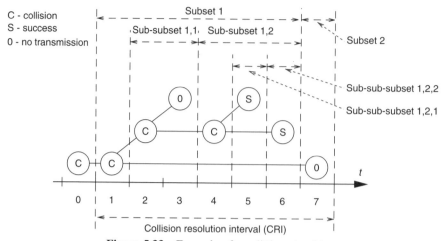

Figure 5.33 Example of a splitting algorithm

In the example presented in Fig. 5.33, the stations from the first subset will try to transmit the packets, and the stations from the second subset wait. If a new collision occurs in the first subset, the subset is furthermore divided into two sub-subsets. This procedure is carried out until all the collisions of a subset are resolved. Afterward, the same procedure is applied to the second subset.

The result of the splitting resolution algorithms is also a stabilization of the network utilization. During the resolution procedure, there is also a need to transfer feedback information. In this way, the stations involved that are informed about the success or collision of sent packets are able to proceed or stop the resolution procedure. However, in a network operating under unfavorable disturbance conditions, such as PLC, there is a higher probability that the feedback information is disturbed that decelerates the resolution process. Finally, the longer collision resolution intervals also increase transmission delays in the network.

5.3.2.3 CSMA Protocol Family

Collision resolution protocols, described above, react on the number of collisions in a network by increasing the contention window (dynamic backoff mechanism), or by starting a mechanism for the collision resolving, or in accordance with current network load it is tried to calculate an optimal transmission probability to reduce the collision probability. A group of MAC protocols with carrier sensing, called Carrier Sense Multiple Access (CSMA) protocols, include another mechanism for the reduction of the collision probability. In accordance with the CSMA, network stations, which have packets to transmit, at first sense the medium to find out if it is already in use by other stations. If this is the case, the sensing stations do not start the transmission and thereby avoid a collision.

Protocol Description
There are two basic sorts of CSMA protocols:

- Nonpersistent CSMA and
- Persistent CSMA, where we usually distinguish between 1-persistent and p-persistent protocol solutions.

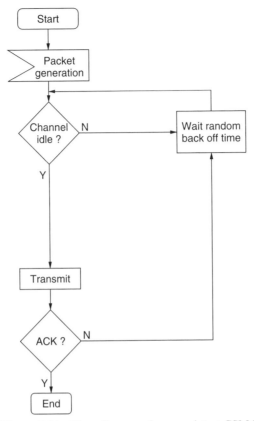

Figure 5.34 Flow diagram of nonpersistent CSMA

In accordance with the nonpersistent CSMA, after a packet is generated, a network station senses the transmission medium, and if it is free the station transmits the packet (Fig. 5.34). If an acknowledgment for the packet is not received after a certain time period, which is necessary for an answer from the receiving station (e.g. base station), the packet has been collided or is lost because of disturbances. In the last case, the disturbances can affect both the packet or the acknowledgment from the base station.

In any case, if there is no acknowledgment, the station has to wait for a random time period to sense the medium again. If a station senses the medium as busy, it becomes backlogged and tries again after a random time as well.

In accordance with the 1-persistent CSMA, after a network station senses the medium busy it continues to sense, and transmits the packet immediately (with the probability 1) after the medium is sensed as free (Fig. 5.35). If the acknowledgment is not received within a designated time period, the station becomes backlogged. After a random time it senses the medium again.

In the case of the p-persistent CSMA, a station senses the transmission medium, such as in nonpersistent and 1-persistent CSMA protocol (Fig. 5.36). After the medium is sensed as free, the station transmits its packet with the probability P, or the station waits a certain time τ with the probability $1 - P$ to sense the medium again. If the transmission system

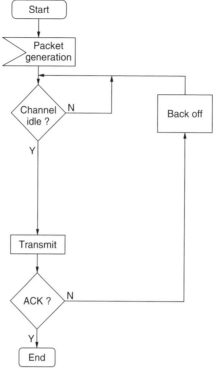

Figure 5.35 Flow diagram of 1-persistent CSMA

is slotted, the station waits for the next time slot or for a number of slots to sense the medium again. After an unsuccessful packet transmission (there is no acknowledgment for the packet), the station becomes backlogged and after a random time it senses the medium again. For $P = 1$, a p-persistent CSMA becomes a 1-persistent CSMA protocol.

Performance Analysis

To evaluate performance of the CSMA protocols, we adopt the same model used for analysis of ALOHA protocols (see above), which is explained in detail in [RomSi90]. All packets transmitted in the network are of the same length T and the maximum propagation delay in the considered network is τ. A normalized propagation time is defined as $a = \tau/T$.

In accordance with CSMA, if a network station starts to send a packet at time t, all other stations will be able to sense the packet after maximum time period of τ. Thus, a collision is possible only if one or more network stations start to send their packets within the time τ (e.g. at moment t', Fig. 5.37). For the general case, we can conclude the following:

- If $t' > t + \tau$, the channel is sensed as busy and no other stations will start to send their packets and no collision will occur, and
- If $t' \leq t + \tau$, the channel is sensed as free because another packet does not yet arrive at the sensing station, and there will be a collision.

PLC MAC Layer

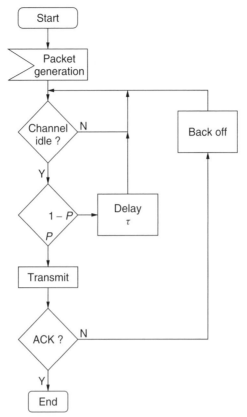

Figure 5.36 Flow diagram of p-persistent CSMA

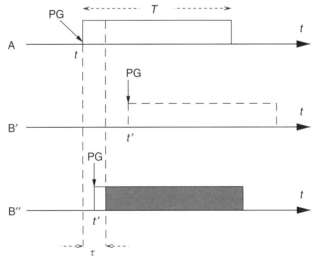

Figure 5.37 Timing diagram for CSMA protocols

In the example presented above, network station A starts to send its packet at the moment t. In the first case, another station (B′) generates a packet at the moment t', after the packet from the station A has already reached all network stations, regarding the maximum propagation delay in the network τ. So, the station senses the medium as busy and prolongs the packet transmission for a later moment and the collision is avoided. In the second case, another station (B″) generates a packet within the interval $[t, \tau]$, senses the medium as free (because the packet from A has not yet reached B) and starts the transmission of the packet, and with it causes a collision.

After mathematical derivation, presented in [RomSi90], network utilization of a nonpersistent CSMA system can be written as

$$S = \frac{Ge^{-aG}}{G(1+2a)+e^{-aG}} \quad (5.32)$$

where G is normalized offered load to the channel, as defined in the analysis of ALOHA protocols, and a is normalized propagation time. Network utilization for different values of the parameter a is presented in Fig. 5.38. It can be recognized that the network utilization in a CSMA system is significantly improved compared to ALOHA protocols (Fig. 5.30). However, the same insatiable behavior of the network utilization still remains.

The performance of the nonpersistent CSMA is improved with lower normalized propagation time a (Fig. 5.38). Thus, in a network with shorter propagation time τ the collision probability is significantly lower, as can be also observed in Fig. 5.37. If the propagation time can be neglected, $a \to 0$, Eq. (5.32) becomes

$$S_{a \to 0} = \frac{G}{G+1} \quad (5.33)$$

So, in this case the network utilization never decreases to zero.

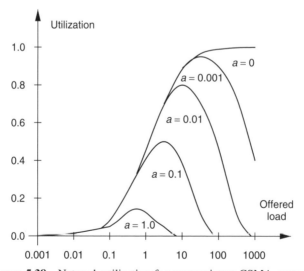

Figure 5.38 Network utilization for nonpersistent CSMA protocol

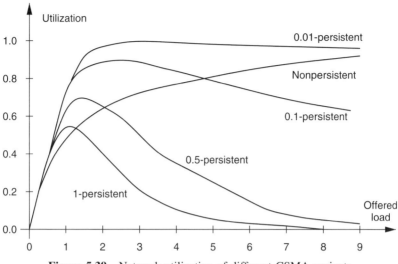

Figure 5.39 Network utilization of different CSMA variants

To improve performance of the nonpersistent CSMA, a further protocols variant, 1-persistent CSMA is introduced. In the case of nonpersistent CSMA (Fig. 5.34), after a station senses the medium as busy, it becomes backlogged and prolongs a packet transmission for a random time. On the other hand, in accordance with the 1-persistent CSMA (Fig. 5.35), after a station senses the medium as busy, it continues to sense the medium and immediately after the medium is free, it starts to transmits its packet. Therefore, the 1-persistent CSMA has an advantage in a lightly loaded network and achieves better network utilization than the nonpersistent CSMA, as shown in Fig. 5.39 [Tane98]. However, the prolongation of the sensing function, and with it the prolongation of the packet transmission after the medium has been sensed as busy, provided by the nonpersistent CSMA, has an advantage in a highly loaded network. In this case, the 1-persistent CSMA principle acts negatively because the immediate packet transmissions, after the medium has been sensed as free, causes frequent collisions in a highly loaded network and decreases the network utilization. Note that in a highly loaded network, there is a larger number of generated packets at the same time, all of them concurring for the transmission at the same time.

The 1-persistent CSMA can be considered as a special case of the p-persistent CSMA, where the access probability p is fixed to 1. So, if the probability p is set to lower values, the collision probability in the highly loaded networks decreases, causing a better network utilization. Accordingly, by setting the access probability to very low values (e.g. 0.01), the p-persistent CSMA achieves significantly better network utilization than the nonpersistent CSMA in the entire considered load area (Fig. 5.39). Of course, if the offered network load is further increased, network utilization achieved by the p-persistent CSMA will decrease under the values achieved by the nonpersistent CSMA. However, in this case the network is extremely overloaded, and therefore is not interesting for applications in real communications networks.

All CSMA protocols can also be implemented as slotted protocol solutions in the same way as is done in the ALOHA protocol. However, the gain achieved in slotted

CSMA systems is very small, as shown in [RomSi90]. Generally, it can be concluded that the CSMA protocols are suitable for applications in short networks where the signal propagation delay is much shorter than the packet transmission time. If the propagation delay is so short as to be neglected, the nonpersistent CSMA can achieve a near-to-full network utilization (Fig. 5.38). However, if the normalized offered load is lower or near to 1 (corresponding to the maximum network data rate), the utilization does not exceed 50%, which is not efficient.

Limitations of CSMA Protocols in PLC Environment
PLC access networks have a centralized communications structure, as mentioned in Sec. 3.1. Accordingly, the communication between subscribers of a PLC network, as well as between PLC subscribers and WAN is carried out via a PLC base station. Therefore, every PLC terminal connected to an access network has to be able to reach the base station by its communications signal. On the other hand, because of the physical structure of a low-voltage power supply network (Fig. 5.40), two PLC modems do not have to be able to reach each other.

If we consider two distant PLC modems in a network, it can happen that the signal transmitted from two terminals A and B reaches the base station, but these two terminals are not able to reach each other directly. This phenomenon, the so-called "hidden terminals problem", is well known from other communications systems, such as wireless networks. In contest of the sensing function provided by a CSMA protocol, it means that if terminal A transmits a data packet, terminal B is not able to recognize it. Consequently, it can sense the medium as free and start transmission of an own packet, causing a collision. Accordingly, the sensing function of CSMA protocols can fail, in particular, cases that decrease the network performance.

Additionally, in networks with unfavorable disturbance conditions, such as PLC, transmitted signals in different network segments can be differently affected by the disturbances (selective disturbances, see Sec. 3.4.4). So, a PLC terminal (e.g. terminal C, Fig. 5.40) can be unable to sense the medium correctly, which can cause an irregular medium access followed by unwanted packet collisions or inefficient transmission gaps.

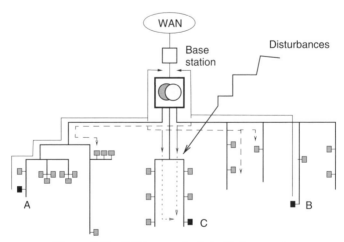

Figure 5.40 Hidden terminals in PLC networks

Protocol Extensions

An improvement of CSMA protocols can be realized by the implementation of the Collision Detection (CD) mechanism that builds a CSMA/CD protocol. This protocol is also specified in IEEE 802.3 standard used in Ethernet LAN systems [Tane98]. The CD mechanism is implemented for the collision detection shortly after it occurs. In this case, the affected transmissions are aborted promptly, minimizing the lengths of the unsuccessful periods. In accordance with the CSMA/CD, if a network station that transmits the data recognizes a collision, it sends a so-called "jam signal" to other stations informing them about the collision. All other stations, which have already started a transmission, interrupt it immediately after the reception of the jam signal. In this way, the occurred collision has the smallest possible influence on the network performance.

The application of the collision detection stabilizes CSMA protocol in the high network load, preventing rapid performance decrease, as shown in [Chan00, RomSi90]. However, because of the hidden terminal problem, described above (Fig. 5.40), it can happen that the jam signal produced by a PLC terminal does not reach every network segment, which reduces the effectiveness of the collision detection function, provided by the CSMA/CD protocol. Additionally, for realization of the CSMA/CD, the transreceivers have to be able to monitor the medium also while transmitting, which increases complexity of PLC modems.

A further variant of CSMA protocol is its combination with the dynamic backoff mechanism for the collision avoidance, described above (Fig. 5.32), forming a CSMA/CA (CSMA with Collision Avoidance) access protocol [NatkPa00, DoufAr02, TayCh01]. This protocol uses an exponential backoff mechanism with the aim to stabilize the network performance. As is already discussed for the collision resolution methods, the result of their application within CSMA protocols is a slight performance improvement and stabilization of the network utilization in highly loaded networks. Some variants of the CSMA/CA are considered for application in PLC networks (e.g. [LangSt00]) and they are also implemented in several currently available commercial products. They usually apply variants of IEEE 802.11 MAC Protocol, described in Sec. 5.3.4, which is based on the CSMA/CA protocol.

ISMA Protocols

As is discussed above, the sensing function provided by the CSMA protocols can fail because of the hidden terminal problem, which exists in PLC and some other communications systems. An Inhibit Sense Multiple Access (ISMA) protocol is proposed for application in wireless networks to deal with the hidden terminal phenomenon [Pras98]. In accordance with the ISMA protocol, a central network instance (e.g. PLC base station) observes status of the uplink transmission channel and informs the network stations about it via a broadcast channel. Thus, the stations that are not able to sense other network stations to estimate if the channel is free or busy receive this information directly from the base station. In this way, the hidden terminal problem is solved, the collision probability in the network is reduced and the network performance is improved.

An ISMA protocol can also be considered as an extended realization of CSMA, where the sensing function is extended by the inhibit sensing, realized by the broadcast information about the channel status. Accordingly, ISMA can be realized as nonpersistent, 1-persistent, as well as *p*-persistent protocol. Furthermore, the ISMA protocols can be implemented as slotted protocol solutions. Various protocol extensions, such as collision detection (ISMA/CD) and collision avoidance (ISMA/CA), can be applied as well. Generally, various ISMA protocols achieve the same performance as corresponding CSMA solutions if the hidden terminal problem is negligible, such is the case in the analysis of various CSMA variants presented above.

5.3.2.4 Collision Elimination Protocols

The sensing function of the CSMA protocol avoids interruption of an existing transmission by network stations that simultaneously have new packets to transmit. However, there is still a probability that more than one station could start the transmission at the same time because of the signal propagation time in the network, causing a collision. A further decrease of the collision probability can be ensured by the application of elimination algorithms, which try to sort out as many stations as possible, before a transmission is started. Such an algorithm is provided by Elimination Yield-Non-Preemptive Priority Multiple Access (EY-NPMA) scheme, which is applied within the channel access control sublayer of HIPERLAN standard for WLAN [Walke99].

According to the EY-NPMA, the channel access takes place in three steps: priority, contention and transmission phases (Fig. 5.41). The contention phase is subdivided into an elimination phase and a yield phase. During the priority phase, a network station with data to transmit senses the channel for a certain number of priority slots (PS). The stations with a lower priority have to sense a higher number of priority slots. If network stations with higher priorities do not compete for access, then the channel is free at this time. After it is recognized by an active station, it sends a burst until the end of the priority phase and can then participate in the contention phase. The burst is a signal for the stations with the lower priority that indicates that the channel is already occupied.

In the contention phase, the network stations, which passed the priority phase and accordingly belong to the same priority class, send so-called "elimination bursts". The burst lengths are variable and correspond to an integer number of elimination slots (ES). The number of elimination slots to be covered by a burst is a geometrically distributed random variable. After the burst, the station observes the channel for the duration of the

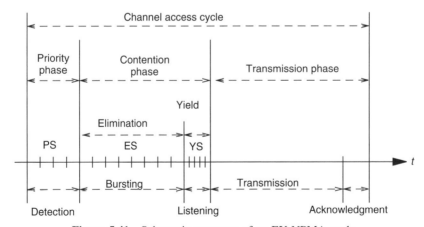

Figure 5.41 Schematic sequence of an EY-NPMA cycle

elimination phase. If the channel is free during this period, the station continues with the yield phase. Otherwise, it does not pass the elimination phase and has to wait for a new cycle to compete again for the transmission.

In the yield phase, each station listens on the channel for the duration of a number of yield slots (YS, Fig. 5.41). The number of slots to be listened to by a station is also a geometrically distributed random variable. If the channel is free during this time, the station starts the transmission immediately. Otherwise, another transmission has already started and the station has to wait for the next cycle to attempt the channel.

The elimination procedure provided by the EY-NPMA ensures a very low collision probability [JancWo00]. However, the maximum network utilization is 60 to 70%, which is achieved if large packets are transmitted (e.g. 1500 bytes). If the transmitted packets are smaller (e.g. ATM cells), the utilization decreases rapidly. A further disadvantage of the EY-NMPA is a relatively large overhead, needed for the elimination procedure that has to be carried out for each of transmitted packets. Additionally, the disturbances, which are presented in PLC networks can also negatively influence the elimination procedure and prolong the time needed for the channel access (e.g. causing irregular elimination bursts).

5.3.2.5 Contention Protocols – Conclusions

After the analysis presented above, it can be concluded that contention MAC protocols (protocols with random access) are generally not able to provide a good network utilization. Another disadvantage of the contention protocols is the difficult realization of QoS guarantees. Because of the collisions, the access time increases especially in highly loaded networks, which is not suitable for the time-critical services. Possible realizations of QoS guarantees (e.g. different priority classes that can be provided by EY-NMPA or Bayesian algorithms) are often linked with an increase of the realization complexity. Additionally, the unconventional noise scenario in PLC networks decreases the performance of implemented algorithms for the QoS realization.

5.3.3 Arbitration Protocols

5.3.3.1 Token Passing

In a network applying a token-passing protocol, the network stations exchange so-called "token-messages" (tokens) in a particular order to specify access right to the medium for every station in the network. A station that just received a token has right to access the medium and transmit its data (Fig. 5.42). After the transmission is completed, the token is sent to another station in the network, which can carry out its transmission. In this way, each network station has an extra time period, determined by the token message, to send its data and collisions between multiple network stations are not possible, which leads to a collision-free network operation. To avoid a situation where a network station transmits its data for a longer time period and with it obstructs other stations to transmit their data, a limit for an individual transmission can be defined. This can be done by limitation of the transmission time, or by specification of a maximum amount of data to be transmitted within one token turn.

The most well-known token-passing protocol is Token-Ring, developed for the application in LANs with a ring topology. According to the token-ring protocol [Chan00], the

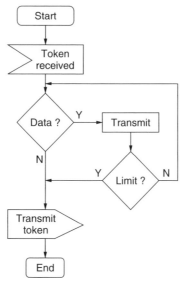

Figure 5.42 Flow diagram of token-passing

transmission rights are specified by the token message that is passed along a ring network between adjacent network stations (Fig. 5.43). When a station has data to send, it waits for a free token to start the transmission. When the station has completed the transmission, it sends a new free token to the next station in the ring. If a network station does not have data to transmit, it passes the token immediately to the next station in the ring.

The token-passing access method is not only applied to the ring networks but also can be used in bus systems (token-bus protocol [JungWa98]), such as the PLC logical bus network (Sec. 3.1.5). In this case, the tokens are not transmitted to a physically adjacent network station, but to a logically adjacent station. The logical order of the stations can be chosen according to their MAC addresses or any other principle.

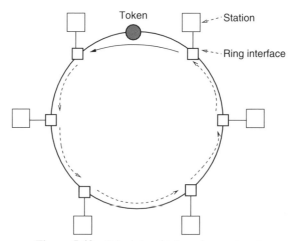

Figure 5.43 Principle of token-ring protocol

A big disadvantage of the token-passing protocols is a possible long round-trip time of the tokens if the number of network stations is relatively high. This increases waiting time for the token and also prolongs the entire transmission delays, which can be inconvenient for time-critical services. On the other hand, if the network is low loaded, the stations with the data to send have to wait for the tokens to start the transmissions, in spite of the fact that the network capacity is mainly unused. In this case, the waiting time, which is independent of the network load is always the same, which is not efficient.

The round-trip time of a token message can be calculated according to the following equation:

$$t_{RTT} = n_{NS}t_T + t_P \tag{5.34}$$

where n_{NS} is number of network stations, t_T is transmission time of a token message, and t_P is transmission time of user packets. In accordance with the token-passing principle, the tokens are exchanged between the stations independently of the traffic in the network; i.a. despite that there are no packets to be transmitted, the tokens are exchanged. Thus, the minimum round-trip time is achieved if there is no data to be transmitted, where $t_{P_{min}} = 0$:

$$t_{RTT_{min}} = n_{NS}t_T \tag{5.35}$$

On the other hand, the maximum round-trip time is achieved when all stations have packets to be transmitted. However, as discussed above (Fig. 5.42), there is a limit for a maximum packet size or maximum transmission time of data from a network station $t_{P_{lim}}$. Accordingly, the maximum round-trip time can be written as

$$t_{RTT_{max}} = n_{NS}t_T + n_{NS}t_{P_{lim}} = n_{NS}(t_T + t_{P_{lim}}) \tag{5.36}$$

In a general case, transmission of a data packet (user packet or token message) consists of the following three delays:

- Delay due to station latency t_L,
- Propagation delay over network segments t_{prop} (e.g. between two neighboring stations), and
- Transfer time of the data contents t_T (all bits of a data packet):

$$t = t_L + t_{prop} + t_T \tag{5.37}$$

The latency and the propagation delays are equal for both tokens and user packets, independent of their sizes. On the other hand, the transfer delay depends directly on the packet (or token) size and available data rate in the network, and is calculated as

$$t_T = \frac{PacketSize}{DataRate} \tag{5.38}$$

Token-passing networks have a distributed access organization due to the token exchange between network stations themselves without a central control unit. On the other hand, PLC access networks have a logical bus topology with a base station at the head of the bus (Fig. 5.26) and are more suitable for access protocols with a centralized approach. The distributed structure of token-passing protocols is also not suitable for application in PLC because of the disturbances, which can often interrupt the token exchange procedure. In this case, the token exchange has to be reset, causing longer transmission delays as well.

5.3.3.2 Polling

As opposed to the token-passing principle, polling is a centralized access method providing a main station to control the multiple access to the shared medium [Chan00]. The base station (e.g. base station of a PLC access network) sends a so-called "polling message" to each network station in accordance with the round-robin procedure or any other cyclic order. If a station receives a polling message, it can transmit the data for a predefined time period (Fig. 5.44). In the case that a polled station does not have data to transmit, it sends a kind of acknowledgment to the base station to inform it that there is no data to send. Afterward, the base station polls the next station in the cycle. The network station transmits also an acknowledgment after the end of a packet transmission, also informing the base station that the transmission is completed (e.g. before a limit is reached) and that the next station in the cycle can be polled.

The polling access procedure can be applied to any network topology. Independent of the physical network structure (e.g. three, bus, ring, Fig. 5.45, or a star network that is typical for wireless communications systems), the polling cycle is carried out in accordance with a logical order of the network stations. Thus, the polling procedure is always carried out from the logically first $(S-1)$ station to the last one $(S-n)$, and so on. Of

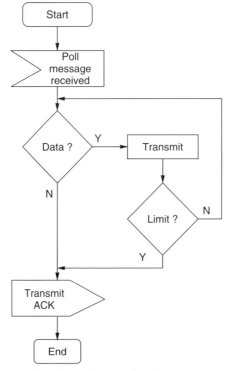

Figure 5.44 Flow diagram of polling access method

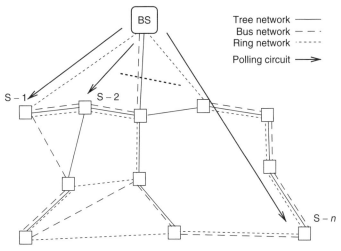

Figure 5.45 Structure of a polling network

course, the propagation time of the polling messages and the data packets transmitted between the base and the network stations depend directly on the physical path used for the transmission.

An advantage of the polling access method compared with token-passing is its higher robustness against disturbances and failures of the network stations. In both cases, the base station, which fully controls the network, can change the polling order to react, in accordance with the current network state. However, with the increasing number of network stations, the round-trip times of the polling messages become longer, similar to round-trip time in token-passing networks. This causes unacceptable waiting delays and low efficiency in the low network load area as well.

The round-trip time of a polling message can be calculated in accordance with the following equation:

$$t_{\text{RTT}} = t_{\text{poll}} + n_{\text{NS}}(t_{L_S} + t_{L_B}) + t_{\text{ack}} + t_P \qquad (5.39)$$

Where t_{poll} is the time needed for transmission of the polling messages to all network stations, n_{NS} is number of stations in the network, t_{L_S} is latency time of a network station, t_{L_B} is the latency time of the base station, t_{ack} is the transmission time of acknowledgments, and t_P is transmission time of packets. The transmission time of a polling message, an acknowledgment or a packet is a function of the propagation delay t_{prop} and the transfer delay t_T

$$t = t_{\text{prop}} + t_T \qquad (5.40)$$

where the propagation time depends on the distance between the base stations and a polled network station and the transfer delay is calculated in accordance with the size of polling message, acknowledgment or packet as well as with the available data rate in the network (Eq. (5.38)). The size of a polling message and of an acknowledgment can be assumed to be constant. Thus, the transfer time of these messages is always the same. On the other hand, the propagation delay differs from station to station and is calculated by

$$t_{\text{prop}} = t_r l \qquad (5.41)$$

where t_r is relative propagation delay (e.g. in s/m) and l is the length of the transmission path.

The transmission time of the packets t_P (Eq. (5.39)) depends also on the propagation and transfer delays, as formulated in Eqs. (5.40) and (5.41). If we assume that each station always has data to transmit and that the transmitted packets have the maximum (limited) size, the maximum packet transmission time for n_{NS} stations in a network is

$$t_{P_{max}} = n_{NS}(t_{prop} + t_{T_{P_{lim}}}) \tag{5.42}$$

However, the transmission paths between different network stations and the base stations have different lengths that influence the propagation delay. So, substituting Eq. (5.41) in Eq. (5.42), the maximum packet transmission time is

$$t_{P_{max}} = \sum_{i=1}^{n_{NS}} (t_r l_i + t_{T_{P_{lim}}}) \tag{5.43}$$

where l_i is distance between the base station and a network station i.

Now, in accordance with the principle of the polling protocol (Fig. 5.44) and Eq. (5.39), we can calculate the maximum round-trip time as

$$t_{RTT_{max}} = t_{poll} + n_{NS}(t_{L_S} + t_{L_B}) + t_{P_{max C}} \tag{5.44},$$

where $t_{P_{max C}}$ represents the time needed for transmission of the packets and the acknowledgments together, so-called "complete packet", because if a station has a packet to transmit, it ends the transmission with an acknowledgment. Therefore, and in accordance with Eq. (5.43), the maximum transmission time of the packets is

$$t_{P_{max}} = \sum_{i=1}^{n_{NS}} (t_r l_i + t_{T_{P_{lim}}} + t_{T_{ack}}) \tag{5.45}$$

where $t_{T_{ack}}$ is transfer delay of an acknowledgment.

The transmission time of the polling messages for n_{NS} stations is

$$t_{poll} = \sum_{i=1}^{n_{NS}} (t_r l_i + t_{T_{poll}}) \tag{5.46}$$

where $t_{T_{poll}}$ is the transfer delay of a polling message.

Finally, by substituting Eqs. (5.45) and (5.46) in Eq. (5.44), the maximum round-trip time of the polling messages is

$$\begin{aligned} RTT_{max} &= \sum_{i=1}^{n_{NS}} (t_r l_i + t_{T_{poll}}) + n_{NS}(t_{L_S} + t_{L_B}) + \sum_{i=1}^{n_{NS}} (t_r l_i + t_{T_{P_{lim}}} + t_{T_{ack}}) \\ &= \sum_{i=1}^{n_{NS}} (2 \cdot t_r l_i + t_{T_{poll}} + t_{T_{P_{lim}}} + t_{T_{ack}}) + n_{NS}(t_{L_S} + t_{L_B}) \end{aligned} \tag{5.47}.$$

The minimum round-trip time of the polling messages is achieved if all network stations do not have any data to transmit during a polling cycle. In this case, the user packets are not transmitted and the packet transmission time is zero. So, by setting $t_P = 0$ in Eq. (5.39), the minimum polling round-trip time is

$$t_{\text{RTT}_{\min}} = \sum_{i=1}^{n_{\text{NS}}} (2 \cdot t_r l_i + t_{T_{\text{poll}}} + t_{T_{\text{ack}}}) + n_{\text{NS}}(t_{L_S} + t_{L_B}) \qquad (5.48)$$

It can be concluded that the delays in a polling system strongly depends on the number of network stations, as well as on their activities, number and size of the transmitted packets. If the round-trip time of the polling messages is too long, this can create problems for realization of time-critical telecommunications services. However, because of the centralized organization of polling access methods, it is possible that the base station polls the network stations according to some predefined priority classes. Thus, the polling principle is applied selectively to different network stations in accordance with their needs, priority levels and QoS requirements to be ensured.

A further possibility to reduce the polling round-trip time is application of so-called "active polling principle" [SharAl01]. The idea of active polling is that only active network stations are polled while other stations are temporarily excluded from the polling cycle. With it, the polling cycle is shortened and the round-trip time of the polling messages is reduced. The active network stations are potential data transmitters and the other stations do not currently send any data. However, the inactive network station has to be able to be added into the polling circuit that can be realized by an extra polling procedure, which is carried out relatively seldom, or by any other registration process.

5.3.3.3 Hybrid Protocols

In accordance with the analysis of contention MAC protocols (Sec. 5.3.2) and the discussions on token-passing and polling protocols presented above, it can be concluded that the behavior of the contention and arbitration access protocols is contradictory; for example, the QoS guarantees can be realized by the polling, but it is not efficient for a higher number of network stations under a relatively low load, where the contention protocol achieves better performance (e.g. 1-persistent CSMA in the network load area, below normalized offered load of $G = 0.5$, as can be seen in Fig. 5.39). Thus, an optimal solution would be a mixed protocol performing as a random access method in low network load area and switching to a dedicated random access method if the network is highly loaded. To combine the features of contention and arbitration protocols, they can be combined in so-called "hybrid MAC protocol solutions" containing both random as well as dedicated access components.

For realization of hybrid MAC protocols, the accessible sections of the network resources provided by a multiple access scheme have to be divided into two groups; one to be accessed in accordance with the random access principle and another that is controlled by a dedicated access mechanism. If we consider a TDMA system usually providing so-called "time frames" that are repeated with a certain frequency (Fig. 5.46) for realization of a hybrid MAC protocols, it is necessary to divide each time frame into two sections with different access principles. In this way, the services with some QoS requirements (e.g. time-critical services such as telephony) can be served during the contention-free phase, which is repeated

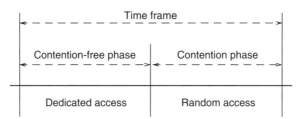

Figure 5.46 Principle of hybrid MAC protocols

in every frame. Other services, without particular QoS requirements (e.g. internet-based data transfer), are served during the contention phase.

The lengths of the contention-free and the contention phases within a frame can be variable, depending on the current traffic conditions. Thus, if there are more connections that require the contention-free access, a longer period can be allocated for this purpose. The association of the network stations to be served within the contention-free period can be carried out during the contention period. So, after a first successful packet transmitted by a network station during the contention phase, for example, using a random access protocol, the station is then associated in the contention-free phase, where it is polled in every time frame. The same principle can be applied for realization of the active polling (described above) for association of new active station in the polling cycle.

Hybrid protocols improve network performance and ensure realization of various telecommunications services with different QoS requirements. However, general problems of the contention and the arbitration protocols (low network utilization in highly loaded networks and long round-trip times and inefficiency in low loaded networks, respectively) still remain in both contention and contention-free protocol phases.

5.3.3.4 Reservation Protocols

In the case of reservation MAC protocols, a kind of reservation of the transmission capacity is done for a particular user or a service. For this purpose, a part of the transmission resource is reserved for realization of the reservation procedure, so-called "signaling". Thus, in a general case, a number of accessible sections of the transmission resources, provided by a multiple access scheme, is allocated for signaling that includes transmission of the user requests (demands) to a central network unit (e.g. PLC base station) and acknowledgments/transmission rights from the base stations. After the reservation procedure is finished, the base station has already allocated necessary network resources for the requested transmission, ensuring a contention-free data transmission.

For realization of reservation MAC protocol in a TDMA system, the time frames are divided into two intervals (Fig. 5.47); one provided for the reservation procedure (R)

Figure 5.47 Principle of reservation MAC protocols

and another for the collision-free data transmission (T). A reservation completed within the request phase of a time frame (e.g. time frame $T - 1$) can affect the same or the next time frame, where the transmission is carried out within time frame $T - 2$, or the transmission can be carried out in any of the next time frames $(T - i)$. Thus, the base station has an opportunity to schedule multiple requests received from different users for various services in accordance with the required QoS, priorities, and so on. The lengths of the reservation and transmission periods within a time frame can be fixed or they can be dynamically changed, depending on the current load situation in a network.

Networks using reservation MAC protocols are suitable for carrying hybrid traffic (mix of traffic types caused by various services) with variable transmission rates [AkyiMc99]. The reservation MAC protocols also ensure realization of various QoS guarantees and achieve a good network utilization as well. So, the first two conditions for an efficient MAC protocol to be applied in PLC access networks are met by the reservation protocols. The application of reservation protocols in networks with a centralized structure, such as PLC access networks with a central base station, seems also to be a reasonable solution. The centralized network organization using reservation protocols can be also seen as a suitable structure for resolving possible irregular situations in the network caused by the disturbances.

The reservation MAC protocols are also proposed for application in broadband PLC networks in [HrasHa01a] and they are implemented by some manufacturers of the PLC equipment as well. Therefore, we present a detailed description of reservation MAC protocols and a comprehensive performance analysis for their application in PLC networks separately in Chapter 6.

5.3.3.5 Arbitration Protocols – Conclusions

Arbitration MAC protocols can provide QoS guarantees for particular services. However, if the number of network stations is large, the round-trip time of token or polling messages increases, which is not suitable for the realization of time-critical services. Both the token-passing and the polling protocols are also not efficient under low network load, if the number of stations is high. On the other hand, reservation MAC protocols are suitable for carrying hybrid traffic, to ensure realization of various QoS guarantees, and achieve a good network utilization. Additionally, the reservation protocols are suitable for implementation in networks with a centralized communications structure, such as in PLC access networks, and make possible an easier realization of the fault management in the network. For these reasons, the reservation MAC protocols can be outlined as good candidates for application in broadband PLC access networks.

5.3.4 IEEE 802.11 MAC Protocol

Originally, the MAC Protocol specified in IEEE 802.11 standard has been developed for application in wireless local area network (WLAN). However, this MAC protocol, in its realization variations, is also applied in broadband PLC access networks by several manufacturers. Therefore, it is important to consider this protocol solution as an actual realization of the MAC protocol for PLC networks. On the other hand, the IEEE 802.11 MAC solution consists of a mix of various protocols, described in Sec. 5.3.2 and

Sec. 5.3.3. Thus, this protocol solution presents in practice a very often used realization of a hybrid MAC protocol implementing different methods for the medium access.

The IEEE 802.11 MAC protocol includes the following three access mechanisms:

- Distributed Coordination Function (DCF),
- Request-to-Send/Clear-to-Send (RTS/CTS) mechanism, and
- Point Coordination Function (PCF).

In the description below, we present the operation of these protocol functions.

5.3.4.1 Distributed Coordination Function

The distribution coordination function of the IEEE 802.11 MAC protocol is based on the CSMA/CA access method. The CSMA is applied in its nonpersistent form (Fig. 5.34), and the collision avoidance is implemented in accordance with the dynamic backoff mechanism (Fig. 5.32). Both methods are described and analyzed in Sec. 5.3.2.

In accordance with the IEEE 802.11 protocol, after a station generates a packet to be transmitted, it waits for the beginning of a new slot (regarding a slot time for synchronization in the network) to sense medium for a predefined sensing time, as is presented in Fig. 5.48. For example, in a WLAN the sensing time is specified by so-called "distributed coordination function interframe space" [Walke99]. If the transmission medium is free during the sensing time, the station transmits the packet. Otherwise, if the medium is busy or partly busy during the sensing period, the station becomes backlogged, such is the case in nonpersistent CSMA protocol.

The backoff time, expressed in time slots (wait counter), is calculated randomly within a contention window (CW). The stations continue to sense the transmission medium and for each time slot decrement a wait counter (wait) if the medium is free, or take no action on the counter if the medium is busy. When the wait counter is decremented to one, the sensing procedure from the beginning is repeated again.

If the station does not receive an acknowledgment after transmission of the packet a collision is determined. In this case, the contention window is doubled and the wait value is calculated from the new contention window (Fig. 5.48). Thus, the backoff interval is increased after each collision until it reaches a maximum size CW_{max}. Afterward, the station waits for the decrementation of the wait time to start a new transmission attempt. For every new packet to be transmitted, the contention window is set to a default value CW_{min}.

5.3.4.2 RTS/CTS Mechanism

The IEEE 802.11 MAC protocol also provides a so-called RTS/CTS (Request-To-Send/Clear-To-Send) mechanism to improve the network performances. A station that has a data packet to send at first transmits an RTS to the destination station (Fig. 5.49). The RTS contains information about necessary network resources (i.e. time) required for the packet transmission. The destination station answers with a CTS to inform the source station that it can start the transmission, and also to inform other network stations that the channel is occupied for a certain time period. During this period, other network stations will not attempt to use the channel, thereby ensuring a collision-free data transmission.

PLC MAC Layer

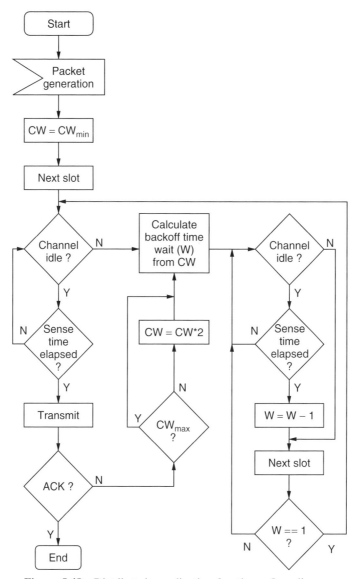

Figure 5.48 Distributed coordination function – flow diagram

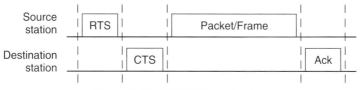

Figure 5.49 RTS/CTS mechanism

The RTS/CTS mechanism has been developed to solve the problem of hidden nodes in wireless networks. Thus, this mechanism serves a kind of virtual carrier sensing, where the sensing function is realized by the exchange of RTS and CTS messages. As is mentioned above, the collisions between user packets are not possible and they can occur only between RTS packets. Since the RTS/CTS packets are relatively small, the waste of the bandwidth due to the collision is significantly reduced, compared to the IEEE 802.11 protocol without RTS/CTS mechanism. However, it is not efficient to apply this mechanism if relatively small data packets are transmitted because of the overhead that originates from the RTS/CTS procedure. Therefore, an RTS threshold is specified to enable the RTS/CTS mechanism for data packets larger than the threshold and to disable it for smaller packets [Walke99]. Note that implementation of the RTS/CTS mechanism within IEEE 802.11 protocol is optional.

The RTS/CTS mechanism presents a realization of a signaling procedure that is typical for reservation access method (Sec. 5.3.3). Accordingly, the IEEE 802.11 MAC protocol applying the RTS/CTS mechanism ensures a nearly full network utilization and realization of particular QoS guarantees.

5.3.4.3 Point Coordination Function

For realization of the point coordination function, the IEEE 802.11 MAC layer frame is divided into two parts (Fig. 5.50); a contention-free phase and a contention phase. The distributed coordination function uses the contention phase for random access to the medium, which is organized according to the CSMA/CA protocol, described above, and the point coordination function, as an optional function of the IEEE 802.11 protocol, is carried out during the contention-free phase. The contention-free phase is used for the application of dedicated access realized by the point coordination function, according to the polling protocol (Sec. 5.3.3). In this way, the services with some QoS requirements (e.g. time-critical services such as telephony) can be served during the contention-free phase, which is repeated in every frame. Other services, without particular QoS requirements (e.g. internet-based data transfer), are served during the contention phase. Thus, implementation of the point coordination function in the IEEE 802.11 protocol presents an example for application of a hybrid MAC protocol.

The lengths of the contention-free and the contention phases within a frame is variable, depending on the current traffic conditions. If there are more connections needing the point coordination function, a longer period can be allocated for this purpose. The association of the network stations to be served within the contention-free period is carried out during the contention period [GanzPh01]. So, after a first successful packet transmitted by a network station during the contention phase, the station is then associated in the contention-free phase, and it is polled in every frame repetition interval.

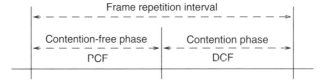

Figure 5.50 Realization of point coordination function

5.4 Traffic Control

Multiple access schemes applied to a broadband PLC network, as is discussed in Sec. 5.2, provide so-called "accessible sections" of the network resources, which are used by multiple network users applying various telecommunications services. An optimized multiple access scheme establishes a basis for an efficient network operation, ensures required conditions for realization of various QoS guarantees, and makes possible implementation of effective error-handling mechanisms. On the other hand, the resource-sharing strategies, or MAC protocols analyzed in Sec. 5.3, manage the accessible sections of the network resources provided by the multiple access scheme. The requirements on the MAC protocols are to achieve a good network utilization, possibly a maximum utilization, to ensure a fast medium access making possible implementation of the services with higher and time-critical QoS requirements, and to be robust against unfavorable disturbance conditions expected in the PLC networks.

For these reasons, the choice of a multiple access scheme and of a MAC protocol is an important issue in the design of a broadband PLC system. However, besides optimization of these procedures, the effectiveness of the network operation and the level of the provided QoS in the network can be significantly improved by implementation of additional mechanisms for traffic control. We divide the traffic control mechanisms in the following three groups:

- Duplex mode (Sec. 5.4.1), as a part of the MAC layers, those optimization can improve network utilization,
- Traffic scheduling (Sec. 5.4.2), representing additional mechanisms to be implemented in the MAC layer to improve QoS in the network, and
- Connection Admission Control mechanism, operating above the MAC layer to secure QoS level in the network (Sec. 5.4.3).

5.4.1 Duplex Mode

5.4.1.1 Principles

The duplex mode defines the organization of traffic in downlink and uplink transmission directions, that is, transmission of data from a base station to network stations and in the opposite direction from the network stations to the base station, respectively. For this purpose, the accessible sections of the transmission resources, provided by a multiple access scheme (Sec. 5.2), are divided into two groups; one of them serving the transmission in the downlink direction and the other for the uplink transmission. The division of the network resources between the downlink and the uplink can be made in two ways:

- FDD – Frequency Division Duplex, and
- TDD – Time Division Duplex.

In the first case, a frequency range is used for the uplink transmission and another range for the downlink. Thus, if we consider an FDMA system, as presented in Fig. 5.51, a number of the frequency bands (transmission channels, Ch) are allocated for the downlink, and the remaining bands are allocated for the uplink, building an FDMA/FDD transmission system.

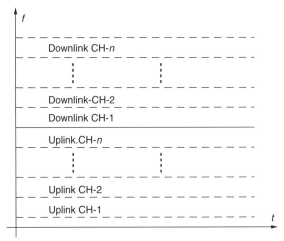

Figure 5.51 FDMA/FDD principle

On the other hand, TDD provides different time frames (TF, Fig. 5.52) where the transmission is carried out by turns in the downlink, or in the uplink. So, in a TDMA system, there are two types of the time frames, downlink and uplink frames, usually containing a number of the time slots (TS), building a TDMA/TDD transmission system.

Besides two combinations of the multiple access schemes and the duplex modes presented above, an FDMA system can be realized with TDD, as well as a TDMA system can be combined with FDD. In the first case, the transmission channel provided by the FDMA scheme are used for both transmission in the uplink and downlink directions. However, there is a division in the time domain, ensuring extra time periods that are used for the uplink and for the downlink transmission (Fig. 5.53).

On the other hand, in case of a TDMA/FDD system, the frequency spectrum is divided in the uplink and in the downlink parts (Fig. 5.54). Thus, the corresponding uplink and downlink time slots are accessed in these two frequency ranges.

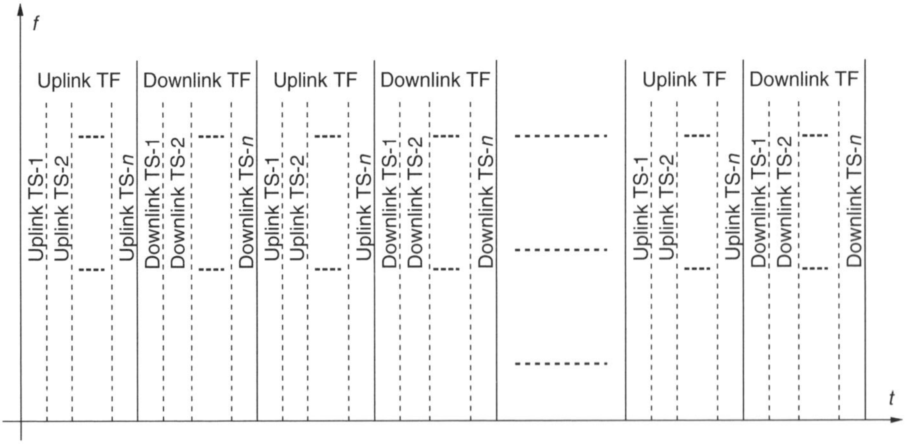

Figure 5.52 TDMA/TDD principle

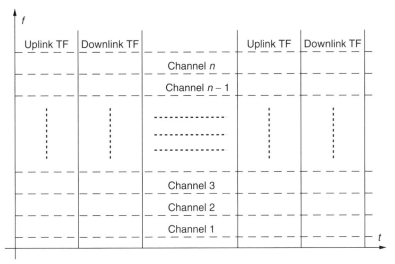

Figure 5.53 FDMA/TDD principle

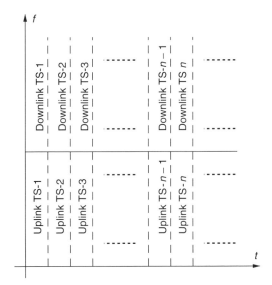

Figure 5.54 TDMA/FDD principle

Both duplex division principles, FDD and TDD, can also be applied to the systems using a CDMA scheme. Thus, in a CDMA/FDD scheme, the transmission channels are provided by the orthogonal codes (Sec. 5.2.3) for both uplink and downlink transmission, but these two transmission directions are realized over separated frequency ranges. On the other hand, in a CDMA/TDD system, the transmission direction is controlled by the turns of uplink and downlink time frames, representing the time periods reserved for each transmission directions. Furthermore, both FDD and TDD can be used in a system based on OFDM access (Sec. 5.2.2). So, in an OFDMA system, the FDD principle can be easily realized by redirecting the transmission channels, provided by the OFDMA scheme, for the transmission in uplink and downlink direction. On the other hand, the TDD principle can be applied to an OFDMA/TDMA system by allocating a number of time slots for each of the transmission directions.

For realization of an FDD system, it is necessary that the network and the base stations are able to operate in two frequency ranges simultaneously; in one range to receive data, and in another one to transmit the data. This increases the complexity of the equipment used in an FDD system. Therefore, the most modern communications systems, especially using a dynamic duplex mode (see below), are organized in accordance with TDD [Rayc99, KellWa99]. However, in this case, the transmission system has to be precisely synchronized.

5.4.1.2 Division Strategies

The duplex mode can be organized as follows:

- Symmetric duplex mode, providing the same data rates for both transmission directions
- Fixed asymmetric mode, providing different but fixed data rates for uplink and downlink
- Dynamic duplex mode, with variable data rates in both directions.

The symmetric duplex mode can be found very often in various communications networks. However, it is not suitable for the access networks, such as PLC, with dominant internet traffic. In this case, the traffic load in the downlink transmission direction is significantly higher than in the uplink direction because of the fact that the subscribers transmit mostly smaller files or requests for the internet pages and download larger files from internet servers, which are usually not placed in the access network and are situated somewhere in WAN (Fig. 5.26).

Therefore, to achieve a better network utilization, the solutions with different data rates in downlink and uplink are more adequate for the PLC access network. As mentioned above, the different data rates can be fixed (e.g. 10% uplink, 90% downlink), or the division of the downlink and uplink transmission capacity is managed dynamically depending on the traffic situation in the network. In the case of fixed asymmetric mode, a higher number of the accessible sections of the transmission resources is allocated for the downlink. Thus, in an FDD system there are more transmission channels allocated for the downlink (FDMA/FDD, Fig. 5.51), or there is a wider frequency spectrum available for the downlink (TDMA/FDD, Fig. 5.54). In the case of TDD solutions, the time frame allocated for the downlink are longer in both FDMA/TDD (Fig. 5.53) and TDMA/TDD (Fig. 5.52) duplex principles. The fixed asymmetric node improves the network utilization and provides higher data rates in the downlink, ensuring enough transmission resources for some specific services (e.g. downloading of streaming data). However, the dynamic duplex mode is more efficient.

5.4.1.3 Dynamic Duplex Mode

In a network with dynamic duplex mode, there is a central unit which controls division of the network resources between uplink and downlink transmission directions. In a PLC environment, we consider a centralized network structure with a base station in its center (Sec. 3.1.5). In this structure, all internal traffic between subscribers within a PLC access network, as well as the traffic between the network access and WAN is carried out over PLC base station. Thus, the base station is able to observe traffic load in both downlink and uplink transmission directions, and accordingly it has the best position in the network to control the dynamic duplex division. Various algorithms for realization of the dynamic duplex mode are proposed for application in different communications system and they are described in numerous publications. In the description below, we outline the main emphasis of these algorithms and discuss the influence of their implementation on the MAC protocol that are considered to be applied in PLC networks.

In accordance with the traffic characteristics in the access networks, such as PLC, where the downlink is expected to be more loaded than the uplink, various dynamic duplex algorithms provide initially higher data rates for the downlink. Thus, in a dynamic TDD algorithm proposed in [ChoiSh96], achieving good results compared to other strategies, the downlink can use more than one half of available time slots if the traffic in the uplink direction is low. Otherwise, the downlink can use a half of the transmission resources.

The traffic load in the uplink direction seems to be significantly lower than in the downlink. However, a good data throughput in the network uplink is also important, especially because of realization of telecommunications services with higher QoS requirements. So, if we consider a time-critical service, an uplink with significantly low data rate can cause unwanted packet delays. Particularly in a network applying a contention MAC protocol, it is important to ensure enough transmission resources in the uplink to reduce the collision probability and following performance decrease. An optimization of the dynamic duplex mode in the case of contention protocols can be carried out in accordance with number of backlogged stations in the network, or with number of the packets to be currently transmitted in the uplink. However, as already discussed in Sec. 5.3.2, an exact estimation of the number of backlogged is complex, causes an additional signaling load in the network and is not suitable for application in networks operating under unfavorable disturbance conditions, such as PLC networks.

On the other hand, the arbitration MAC protocols are more suitable to be combined with the dynamic duplex mode. Thus, in the case of a polling system, the base station can always have an exact information about expected traffic load in the uplink, and accordingly it can take necessary actions to control the duplex mode in an optimal way. Particularly, reservation MAC protocol offer an opportunity for the efficient application of the dynamic duplex mode. In this case, the base station has the information about expected traffic in the uplink immediately after it has received transmission requests from the network stations (Fig. 5.47). So, the optimal division of the transmission resources between the downlink and the uplink can be carried out before the actual transmission is started.

5.4.2 Traffic Scheduling

Mechanisms for the traffic scheduling in communications systems are responsible for management of different data flows transmitted through a network with respect to the fulfillment of the required QoS guarantees for particular telecommunications services. There are a large number of mechanisms for the traffic scheduling investigated for implementation in various telecommunications technologies (ATM, modern IP networks, etc.). Thus, the mechanisms, such as call admission control, traffic shaping and policing, different methods for congestion control in a network, and so on, are proposed for implementation in ATM systems [Onvu95], and they are considered in numerous application variations for the usage in other telecommunications networks as well.

However, we limit the presentation of the traffic scheduling mechanisms on several disciplines that are important in contents of MAC protocols and their application in broadband PLC access networks. In the description below, we analyze implementation possibilities of priority realization, QoS control mechanisms and fairness provision in contention and arbitration MAC protocols. The CAC mechanisms are analyzed separately in Sec. 5.4.3.

5.4.2.1 Priority Realization

In telecommunications systems, a particular priority level can be specified to a network user, a service or a connection. If we consider various telecommunications services divided in certain service classes with assigned priorities, they are treated by a communications system in accordance with their priority levels. Thus, the services belonging to a class with a higher priority level are served before the services with lower priorities. The same principle is used if the prioritizing is applied to distinguish between different user or connection classes.

In contention MAC protocols (Sec. 5.3.2), there is no guarantee that a user packet will be successfully transmitted with a certain time period because of possible collisions. Therefore, it is difficult to realize priorities in networks with the random medium access. However, there is a possibility to distinguish between different priority classes by assigning them different access probabilities. Thus, in an algorithm based on Pseudo-Bayesian stabilization protocol, presented in [FrigLe01, FrigLe01a], services with a higher priority level have a higher access probability. This principle can be applied in any contention-based MAC protocol by assigning different contention windows to different services, in accordance with the priority levels. So, the backoff retransmission time for a service with a higher priority (e.g. priority 1, Fig. 5.55) is calculated from a smaller contention window than is the case for services with lower priorities. If we assume that $CW_1 = nCW_n$, the access probability for the priority class 1 is n times higher than the access probability for the priority class n. The same principle can be used if a dynamic backoff mechanism is applied, as well as for the transmission probability after sensing period in a p-persistent CSMA protocol (Sec. 5.3.2).

A further possibility for the realization of priorities in networks using contention protocols is given in elimination protocols (Sec. 5.3.2). In this case, the services (or users) with a lower priority level are sorted out as first during the collision elimination phase. Thus, the services with higher priority are always advantageous compared to the low-priority

Figure 5.55 Realization of priorities in contention MAC protocols

Figure 5.56 Protocol phases in a reservation access method

services. The collisions are possible only between services belonging to a same priority class.

As opposed to the contention MAC protocols, realization of the priority in the arbitration protocols is significantly easier. An example of a priority realization is selective polling (Sec. 5.3.3), which can be implemented to poll the network station belonging to a higher priority class more frequently than the stations with lower priorities. The polling procedure is fully controlled by the base station, and accordingly strict guarantees for the duration of a polling cycle can be given for each priority class (e.g. user priority class).

As is already mentioned in Sec. 5.3.3, the reservation protocols contains reservation and transmission phases. During the reservation phase, network stations demand usage of various services by sending transmission requests to the base station, whereas the data transfer is carried out during the transmission phase (Fig. 5.56). The time between these two protocol phases is used for scheduling of the transmission requests, where the base station has a convenient opportunity to reschedule the transmission requests in accordance with the assigned priority level of the requesting stations or services. Thus, the priority implementation in a network using the reservation access methods seems to be easy.

For example, a reservation is done during the time slot i. However, the transmission of user packet is scheduled for the time slot $i + 3$. In between, during time slots $i + 1$ and $i + 2$, transmission of other requests and packets is carried out. So, if a request for a service with a higher priority arrives, the transmission originally scheduled for the time slot $i + 3$ is postponed for later, to ensure an earlier transmission of data packets belonging to a service with a higher priority.

5.4.2.2 QoS Control

Every telecommunications service requires particular QoS guarantees, which have to be kept for the entire duration of a connection. The QoS guarantees in various telecommunications systems can be specified by the following performance parameters:

- Blocking probability – the probability that a subscriber is not able to realize a connection (e.g. telephony connection) or data transfer because there are currently no free network resources.
- Dropping probability – the probability that a connection or data transfer has to be interrupted because of a decrease of the network capacity (e.g. resulting from new incoming connections with a higher priority or occurrence of disturbances).
- Data rate (data throughput) – as the necessary transmission rate for a service.
- Loss probability – the probability that a portion of transferred information is lost during a transmission (e.g. packet loss probability).
- Transmission delay (considered on different network layers) – the time needed for transmission of a data unit (segment, packet, file, etc.).

During operation of a communications system, the network conditions are permanently changed. This is caused by a varying number of subscribers that actively used the network service, by various service using the network resources with a changing intensity, dynamic traffic characteristics of different services, varying activity of individual subscribers, and so on. Particularly in a network operating under unfavorable noise conditions, the available data rate in the network can frequently change in accordance with current disturbance behavior. All these factors directly influence the data transmission in a network and can cause degradation of QoS in the network. To reduce the possible QoS degradation in a network, efficient CAC mechanisms (Sec. 5.4.3) can be implemented to limit the number of admitted connections in the network (e.g. users, various data connections, etc.). However, in spite of the usage of such mechanisms, the QoS degradation for particular services, already admitted in a network, has to be managed by the MAC layer.

It is possible to control data throughput and transmission delays of the connections existing in the network by tuning parameters of the MAC protocols in accordance with the current network conditions. Also, by a control of the transmission delays, it is possible to influence blocking and dropping probability as well as the packet losses. Thus, if the QoS degradation for a particular connection (or user, or service) is observed, this connection has to be preferred until its QoS level becomes satisfied. Of course, the privileged connection must not be carried out to handicap other connections in the network. The temporary preferential treatment of connections with the degraded QoS can be ensured by assigning them to a service class with higher priority for a while. So, the same mechanisms discussed for the contention and the arbitration protocols for the priority realization (described above) can be applied for the QoS control, too.

5.4.2.3 Fairness

As we described above, to ensure QoS guarantees for various telecommunications services in a network, it is possible to divide services, as well as users, in several priority classes. In this case, each priority class is served in accordance with the specified QoS requirements for the class, and with it, is also possible to fulfill the requirements of each individual service or user. However, the traffic patterns caused by various telecommunications services belonging to a same priority class can significantly distinguish. For example, application of a specific service produces relatively high traffic load and another service from the same priority class produces a lower traffic load. The different traffic characteristics of these services can cause so-called "unfairness" where the performance evaluated for each of the services (e.g. data throughput, delays, etc.) significantly differs. The unfairness between services or users can also be caused by other factors; position of a station in the network (e.g. a far station), order of station association in the network (e.g. association in a polling or scheduling cycle), and so on.

The task of a MAC protocol is to manage access of multiple users applying various services to a shared transmission medium. There, the MAC protocols have to ensure a certain fairness between network users and services, which belong to the same priority class. This can be realized in accordance with the same principles that applied for the priority realization and QoS control, as is described above. So, with an appropriate variation of the access probabilities in the contention MAC protocols, as well as with the appropriate scheduling in the arbitration protocols, network performance of the disadvantageous connections can be improved and equalized with other connections from the same priority class.

5.4.3 CAC Mechanism

Since every telecommunications system provides a finite transmission capacity (a maximum available data rate), a network can carry only a limited number of connections simultaneously. Additionally, if the services with higher data rate and QoS requirements are transferred, the transmission limits can be quickly achieved, particularly in networks with limited data rates, such as recent PLC access networks. Therefore, communications networks apply very often call/connection admission control mechanisms (CAC), which limits the number of connections to be admitted in the network in accordance with current QoS level and data rates that can be ensured for individual connections, applying various telecommunications services. The limitation of the number of admitted connections in a network is specified by so-called "admission policy". Additionally, in networks operating under unfavorable noise conditions, such as PLC, the influence of disturbances on the change of the available data rate in the network has to be particularly considered in an applied CAC mechanism as well.

5.4.3.1 Admission Policy and Channel Allocation

The QoS requirements of various traffic classes caused by numerous telecommunications services are different (Sec. 4.4.3). Therefore, an admission policy has to be specified for each traffic class to make a decision if there are enough transmission resources in the network, which can ensure the required QoS. The decision can be made in different ways; for example, as presented in [BeardFr01], according to the current network conditions; free network capacity, current transmission delays in the network, and so on. A possibility for application of separated admission policy for various traffic classes can be ensured by allocation of different logical transmission channels, provided by a multiple access scheme (Sec. 5.2.4), to various traffic classes, as is shown in Fig. 5.57. Besides reserved, idle and error states, the transmission channels can be allocated for different kinds of services that are divided into a number of classes. So, a CAC mechanism and a corresponding admission policy can be implemented separately for each service class, depending on the required QoS guarantees.

The channels to be used by a particular service class can be allocated in a fixed manner, or the allocation can be organized dynamically, depending on the current traffic conditions in the network and priorities of particular service classes. There are numerous proposals for different channel (resource) allocation strategies, which can be classified in following five types [BeardFr01]:

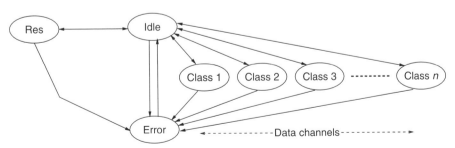

Figure 5.57 Channel state diagram for multiple service classes

- Complete partitioning – where a set of the transmission resources, a number of accessible sections of the network resources (e.g. a number of time slots within a repetition time frame), can be exclusively used only by a traffic class. This method is not efficient if a particular service belonging to traffic class is currently not used, because the exclusively reserved part of the transmission resources only for this class remains unused.
- Guaranteed minimum – allocating a minimum part of the transmission resources for each traffic class, where the remaining network capacity is shared by all traffic classes, for example, in accordance with a complete sharing strategy (see below). In this case, a smaller portion of the transmission resources can remain unused if a traffic class is inactive. However, the allocated minimum capacity for particular classes suffers from the same efficiency problem such as the complete partitioning method.
- Complete sharing – allows that all connections are admitted to use the transmission resources simply if they are available at the time a connection is requested and if they are sufficient to fulfill the required QoS for the requested connections.
- Trunk Reservation – distinguishes between different priority classes of users or services by allowing a particular class to use the transmission resources until a particular part of the resources remains unused. For the classes with lower priority, the defined part of the network resources to remain unused is specified to be higher than for the classes with higher priorities.
- Upper limit policy – where an upper limit on the amount of resources that can be used by a priority class is strictly defined. An upper limit policy provides a threshold for every priority class, and upper limits for the lower priority classes prevent overloads that could affect classes with the higher priority. On the other hand, there is no upper limit for the class with the highest priority. This method clearly handicaps connections belonging to the lower priority classes.

5.4.3.2 CAC in Networks with Disturbances

In communications systems operating under an unfavorable noise scenario, such as PLC, there is a need for the application of a reallocation strategy, making a network more robust against disturbances. Such communications systems are characterized by a stochastic capacity change caused by unpredictable disturbance occurrence [SiwkoRu01]. The conventional CAC policies consider only currently available resources in a network to decide if a new connection will be admitted. However, the disturbances can negatively influence the network operation and decrease the available network capacity, which can lead to dropping (or interruption) of existing connections in the network (already admitted connections).

For many communications applications, dropping of an existing connection after it is already admitted in the network is considered as less desirable than blocking of a new connection to be admitted in the network. Therefore, at admission of new connections in the network, attention has to be payed to the possible future events in the network, caused by the disturbances that possibly decrease the available network capacity. To avoid the interruption of the existing connections, there is a need for a CAC strategy that specifies a spare part of the network capacity that is used for the replacement of disturbed parts of the resources, ensuring continuation of the existing connections (e.g. by providing a number of reserved transmission channels, Fig. 5.57).

The dropping probability cannot be reduced to zero, and therefore there is also a need for definition of so-called "dropping policy" for different service priority classes, specifying a dropping probability that is guaranteed for different services. An admission policy considering the disturbance conditions in PLC networks is proposed in [BegaEr00] and is presented below.

5.4.3.3 A CAC Mechanism for PLC

The performance of a PLC access network depends, among others, on the mix of used telecommunications services, the user behavior, and the available system capacity. In this analysis, we group all services into two different classes, circuit and packet switched. For circuit-switched connections, such as voice, the transmission resources are reserved for the entire duration of the call (Sec. 4.4). For packet-switched connections, the resources are reserved as long as data for transmission are available. Regarding the arrival and service process, state-dependent negative exponential distributed interarrival and service time are assumed for the voice connections. The data traffic is modeled on the burst level, where the bursts arrive in accordance with a Poisson process and burst sizes are assumed to be geometrically distributed.

A PLC network is modeled as a loss system with $C(t)$, as the total number of transmission channels (e.g. with capacity of 64 kbps) available at the time t. Depending on the disturbances, there are 0 to $C^{(max)}$ available channels. If $X_1(t)$ denotes the number of voice calls in the network and $X_2(t)$ the number of data bursts in the system at the time t, then

$$X(t) = (X_1(t), X_2(t), C(t))$$

defines a continuous-time stochastic process with finite discrete state space. The set of allowed states depends on the CAC admission policy, defined for the considered PLC network (Fig. 5.58).

Let $b_2(x)$ define the state-dependent bandwidth in number of transmission channels of one data burst in state x and assume that on average all data bursts get the same bandwidth between 0 and $b_2^{(max)}$, where $b_2^{(max)}$ is the maximum bandwidth, which one data burst can get. On the other hand, let b_1 be the fixed bandwidth of one voice connection, which corresponds to one transmission channel. To introduce flexibility in the resource allocation, the following two minimum bandwidth thresholds are defined:

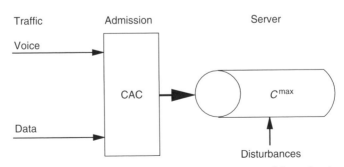

Figure 5.58 Analytical PLC network model on call/burst level

- $b_2^{(\min 1)}$ – minimum bandwidth that data bursts have to reduce to in favor of arrivals of voice calls.
- $b_2^{(\min 2)}$ – minimum bandwidth the data bursts have to reduce to in favor of new data burst arrival, and if it is not possible the new arrival is blocked.

The interarrival and holding time of disturbances are assumed to be negative exponentially distributed in accordance with the noise behavior expected in PLC networks, described in Sec. 3.4. The disturbance can be considered to affect the transmission channels independently or to affect multiple transmission channels. Two values C_1 and C_2 for voice calls and data bursts, respectively, are introduced as the number of reserved channels with respect to these services. Now, we can define an admission policy for the considered PLC networks with respect to voice calls as

$$x_1 + x_2 b_2^{(\min 1)} \leq C(x) - C_1 - b_1 \tag{5.49}$$

This policy can be interpreted so that in state x a new arrival of a voice call is accepted, if after its admission the sum of all minimum bit rates with respect to voice calls is not greater than $C(x) - C_1$, hence the condition presented in Eq. (5.49) must hold. Similarly, the admission policy for data bursts is defined as

$$x_1 + x_2 b_2^{(\min 2)} \leq C(x) - C_2 - b_2^{(\min 2)} \tag{5.50}$$

5.5 Summary

Task of a MAC layer is to manage access of multiple network stations to a shared transmission medium. Functions of a MAC layer can be divided into following three groups: multiple access, resource sharing strategy (MAC protocol), and traffic control functions. The multiple access scheme establishes a method of dividing the transmission resources into accessible sections that can be used by the network station to transfer various types of information. The task of a MAC protocol is the organization of a simultaneous access of the multiple network stations to the accessible sections of the network transmission resources, provided by a multiple access scheme. Traffic control functions, such as dynamic duplex mode, traffic scheduling and connection admission control are additional features of MAC layer and protocols, ensuring realization of particular QoS guarantees in a network and improving the network efficiency.

The MAC layer is a component of the common protocol architecture in every telecommunications system, developed in accordance with the specific features of a communications network and its environment. Broadband PLC access networks are characterized by their specific network topology determined by topology of low-voltage supply networks, features of the power grids used as a transmission medium, operation under unfavorable noise conditions and with relatively limited data rates caused by EMC restrictions, and specific traffic mix to be carried over the network as a consequence of application of various telecommunications services. Thus, a MAC layer to be applied to the PLC networks has to fulfill their specific requirements, which can be summarized as follows:

- Multiple access scheme has to be applicable to the transmission system used for realization of a PLC network, it has to provide realization of various telecommunications

services, and to ensure a certain robustness against unfavorable disturbance conditions in the network.
- MAC protocol has to achieve a good utilization of the limited data rates in PLC networks, to ensure realization of various QoS guarantees for different kinds of telecommunications services, and to operate efficiently under the noise presence as well.

All three basic multiple access schemes (TDMA, FDMA and CDMA) can be applied to the transmission systems, such as spread-spectrum and OFDM-based solutions, which are outlined as suitable solutions for PLC. Because of the requirement for a good network utilization in PLC networks and provision of various QoS guarantees, the segmentation of user packets into smaller data units to be transmitted over the network seems to be a reasonable solution, ensuring a better efficiency of applied error-handling mechanism and providing a finer granularity of the network resources. On the other hand, various FDMA-based solutions, such as OFDMA and OFDMA/TDMA, are especially robust against narrowband disturbances, which are also expected in the PLC networks, and therefore they are considered as suitable schemes for PLC.

Appropriate solution for a MAC protocol to be applied to the PLC networks, and also to other communications systems, can be investigated independently of the applied multiple access scheme by usage of logical channel model. The consideration of different MAC protocols for the uplink of the PLC networks can be summarized as follows:

- Fixed access strategies are not efficient if they carry bursty data traffic, which is expected to be dominant in access networks, such as PLC, and therefore they are also not suitable for application in PLC access networks.
- Dynamic MAC protocols with contention are suitable to carry the bursty traffic, but they do not achieve good network utilization and do not provide an easy realization of QoS guarantees.
- The dynamic protocols with arbitration, such as token passing and polling, can provide realization of various QoS guarantees in some cases, but they can also cause longer transmission times, which is unsuitable for time-critical services.
- Reservation MAC protocols ensure collision-free data transmission, the realization of QoS guarantees and they also provide good network utilization. In the case of reservation protocols, the transmission is controlled by a central unit (base station), which is favorable for realization of an efficient fault management in a centralized network structure, such as PLC. Therefore, the reservation protocols are outlined as a reasonable solution for application in the PLC access networks.

IEEE 802.11 MAC protocol, originally developed for wireless communications networks (e.g. WLAN), is very often applied in various PLC systems. This protocol is based on an access principle with possible contentions between multiple network stations (CSMA/CA). However, additional features of the IEEE 802.11 MAC protocols, which are a combination of the contention and a polling-based contention-free access principle building a hybrid MAC protocol and application of so-called "virtual sensing function", which can be understood as an application of reservation access principle, ensure realization of the required QoS guarantees and provide a good network utilization.

Application of a dynamic duplex mode dividing the available data rates between uplink and downlink transmission directions can significantly improve network efficiency. On the other hand, implementation of traffic scheduling mechanisms within the MAC protocols can be necessary to allow realization of multiple priorities in a network for different user or service classes, to provide a continuous control of realized QoS in the network, as well as to ensure fairness between multiple users or services belonging to a same priority class. Finally, to be able to guarantee the QoS in the network, it is necessary to implement a CAC mechanism, acting above the MAC layer, to restrict the number of connections, subscribers, or service simultaneously using the network resources. An appropriate admission policy for PLC has also to consider possible variations of the available data rate in the network, which are caused by the disturbances.

6

Performance Evaluation of Reservation MAC Protocols

As concluded in Sec. 5.3.3, networks using reservation MAC protocols are suitable for carrying a traffic mix caused by various telecommunications services with variable transmission rates, ensuring realization of various QoS guarantees and achieving good network utilization. On the other hand, the reservation protocols are suitable for application in networks with a centralized structure, such as PLC access networks with a central base station. The centralized network organization that uses reservation protocols is also considered a suitable structure for resolving unusual situations in the network caused by the disturbances. Therefore, we prefer application of the reservation protocols in broadband PLC access networks. Additionally, the RTS/CTS mechanism, implemented within IEEE 802.11 MAC protocol (Sec. 5.3.4), which is applied to several recent PLC systems, can be seen as a reservation access method as well.

For all these reasons, it is necessary to analyze the reservation MAC protocols as regards the contents of their application in PLC networks in more details. At first in this chapter, we describe components of the reservation MAC protocols and make proposals for their implementation in PLC networks (Sec. 6.1). In Sec. 6.2, we present a modeling approach for investigation of signaling MAC protocols, carried out in Sec. 6.3, which results in a proposal for a two-step reservation MAC protocol to be used in broadband PLC access networks. Finally, we consider implementation of various error-handling mechanisms within per-packet reservation MAC protocols (Sec. 6.4) and compare several advanced protocol solutions for PLC, including a discussion of possibilities for the realization of QoS in PLC networks using these protocols (Sec. 6.5).

6.1 Reservation MAC Protocols for PLC

A reservation MAC protocol merges several functions that are necessary for the realization of medium access and the entire signaling procedure between multiple network stations and a base station. To analyze operation of the reservation MAC methods, we define the following four protocol components:

- reservation domain, specifying a data unit or a time period for which the reservation is carried out;

- signaling procedure, describing an order of events for the exchange of signaling messages between the network stations and the base station;
- access control, ensuring collision-free medium access for multiple stations; and
- signaling MAC protocol, applied in the part of the network capacity allocated for realization of the signaling procedure (e.g. signaling channel).

6.1.1 Reservation Domain

According to the procedure of the reservation MAC protocols, a prereservation of network capacity is carried out for a user/subscriber or for a particular service. The reservation can be carried out for the entire duration of a connection or in part for its certain partitions. The chosen reservation domain has a big influence on network performance, especially on network utilization, which is important for transmission systems with limited data rates, such as PLC. In the following section, we present several possibilities for the choice of the reservation domain to be applied within a MAC protocol.

6.1.1.1 Connection Level Reservation

Reservation at the connection level is well known from the classical telephony network. Once a channel is allocated to a voice connection, it remains reserved for the connection until the end of the call. This reservation method is also known as *fixed access strategy*, described in Sec. 5.3.1, which is outlined as not a suitable solution for data transmission with typically bursty traffic characteristics.

The main disadvantage of the call level reservation domain is that the allocated network capacity remains unused during transmission pauses, which very often occur in a data connection (Fig. 5.19). This is not efficient and causes bad network utilization. On the other hand, the bursty characteristic of a data stream can cause so-called transmission peaks, when the capacity of the allocated channel is not enough to serve the data burst causing additional transmission delays and decreasing data throughput.

6.1.1.2 Per-burst Reservation

The per-burst reservation method is very often used for data transmission in wireless networks (e.g. GPRS [KaldMe00]). The reservation is carried out at the beginning of each data burst and the allocated network resources remain reserved for the data burst until its end, which is specified by a time-out period (Fig. 6.1). If there are no new packets within a time-out, the burst is considered as finished and the allocated network resources are free for data bursts from other data users.

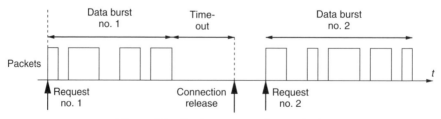

Figure 6.1 Per-burst reservation method

A data burst consists of a number of packets generated by a network station. The packets can be transmitted one after the other, but there can be an interval between the packets. So, during the empty intervals between packets, the allocated network resources remain reserved and this part of the network capacity is not used for any transmission. Accordingly, during a time-out period for the recognition of the end of a data burst, reserved capacity is lost as well. However, per-burst reservation is more efficient than the reservation on the connection level for data traffic that has a dynamic characteristic.

6.1.1.3 Per-packet Reservation

To be able to avoid the transmission gaps between packets, which occur within the per-burst reservation method (Fig. 6.1), the reservation can be carried out for each generated packet (e.g. IP packet). In this case, the transmission gaps that occur during a data connection can be used by other data transmissions, which increases utilization of the common network capacity. However, the per-packet reservation method significantly increases network load caused by the signaling procedure. This is determined by the need for an exchange of signaling messages between network stations and the base station for each transmitted packet.

In Sec. 5.2.1, we mentioned that a segmentation of user packets into smaller data units, the so-called data segments, is useful for improving the performance of networks with limited data rates, such as PLC access networks. Thus, a special case of per-packet reservation method is per-segment reservation, which is applied to some communications protocols (e.g. DQDB [ieee90]). Per-segment reservation can improve the fine granulation of the network capacity, ensuring good network utilization and giving the possibility for realization of various QoS demands provided by the data segmentation. However, the signaling load becomes very high because of the frequent transmission requests and the corresponding acknowledgment packets.

6.1.1.4 Combined Reservation Domains

In accordance with the discussion of the different reservation domains presented above, the choice of an optimal reservation domain depends strongly on the kind of services for which the reservation is carried out; for example, in classical telephony, reservation of a channel for the entire duration of the connection is a reasonable solution. On the other hand, as is shown above, the per-packet solution is good for services with a dynamic characteristic such as data transmission. Therefore, a combination of various reservation principles depending on requested services seems to be a suitable solution for the reservation domain. In this case, a particular reservation domain is applied for each group of telecommunications services, or for each service or traffic class.

For example, if only primary telecommunications services are considered (telephony and Internet, Sec. 4.4.2), the following combination of reservation principles can be specified as an optimal solution: connection level reservation for telephony, and per-packet reservation for Internet-based data transmission. If we consider some advanced data services with higher QoS requirements and stronger delay limits (e.g. video transfer), the per-packet reservation domain can cause a very long reservation procedure, which has to be carried out for each transmitted packet. In this case, the per-burst reservation domain can be a suitable solution, making a compromise between the long signaling delays, caused by the per-packet reservation and inefficient connection level reservation domain.

6.1.2 Signaling Procedure

The signaling procedure specifies an order of events for the exchange of signaling messages between network stations and the base station, which is necessary for realization of the reservation procedure. For a general case, the types of signaling messages that have to be exchanged for the realization of a simple signaling procedure, containing a minimum signaling information, are the following:

- *Transmission request/demand* – sent by network stations to the base station in the uplink transmission direction to request usage of particular services. A request contains information about the requesting station (e.g. ID, priority level, etc.), the requested service (service category or class), and service-specific information (e.g. number of data units/segments to be transmitted).
- *Allocation message* – transmitted by the base station in the downlink after receipt of the request, to inform the network stations about their access rights. An allocation message can contain the following information:
  - allocated transmission channel(s) or time slot(s) to be used for the requested service, and
  - a time or a time slot for beginning the transmission.
- *Acknowledgment* – transmitted by the base station to the network stations to confirm receipt of a transmission request (and also other messages, if any).

Of course, the signaling procedure can contain further types of control messages in a real communications system in accordance with specific implementation and realization requirements in a network.

Acknowledgements and allocation messages can be transmitted separately (e.g. in CPRMA protocol, [AkyiMc99]) or in the same packet. In the first case, the station receives an acknowledgment immediately after the base station receives its transmission request (Fig. 6.2). The acknowledgment informs the requesting stations only that its transmission request has arrived at the base station. The allocation message, containing information about access rights, is transmitted later, directly before the transmission starts. In the second case, both acknowledgment and allocation messages are transmitted jointly, immediately after the transmission request is received by the base station. The transmission of only one control message per request is the more efficient solution because of the following reasons: downlink of the signaling channel(s) is less loaded and error probability for the control messages decreases because there is a lower number of transmitted control messages.

Traffic conditions in the network can change either because of the arrival of connections with higher priorities than currently admitted connections in the network (Sec. 5.4.2), or because of the variation of available data rate in the network caused by disturbances. In both cases, it can happen that connections with lower priorities have to be postponed to ensure an immediate transmission of data from connections with higher priorities. Then, in a network using a reservation MAC protocol with joint control messages, an

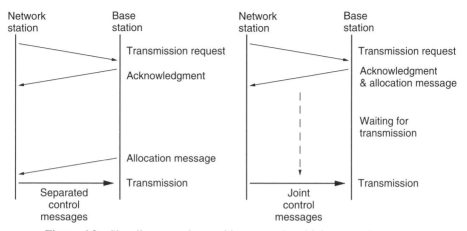

Figure 6.2 Signaling procedures with separated and joint control messages

additional allocation message informing the network stations about the rescheduling of their connections has to be sent by the base station. However, the additional allocation message can be corrupted by the disturbances as well. Additionally, the disturbances can affect a network selectively, causing a group of network stations not to be able to receive the reallocation message at that moment, whereas all other stations that operate under better noise conditions can receive the message. The network stations that did not receive the allocation message or that received an erroneous allocation message are not correctly informed about the rescheduling, which can cause unwanted transmission collisions in the network.

6.1.3 Access Control

6.1.3.1 Access to the Logical Transmission Channels

PLC networks are expected to ensure realization of various telecommunications services. For this purpose, accessible sections of network resources, provided by a multiple access scheme, can be allocated for particular services carrying their data packets, as considered in Sec. 5.4.3. So, in the logical channel structure presented in Fig. 6.3, a transmission channel can be allocated, usually in a dynamic manner, for transmission of various service classes. As mentioned in Sec. 5.3.3, for realization of the reservation procedure within

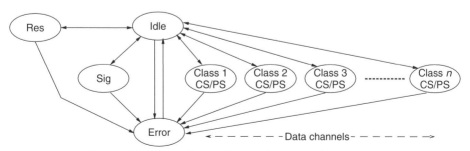

Figure 6.3 Channel state diagram for reservation protocols

a reservation MAC protocol, it is necessary to allocate a certain portion of the network transmission resources (signaling channel, Sig). Thus, a number of the accessible portions of network resources are allocated for signaling, which is carried out between network stations and a base station. The number of the accessible sections used for signaling and their common data rate can be fixed, or the signaling data rate can be variable as well.

Basically, the data channels used for transmission of various service classes can be divided into two types:

- circuit switched (CS), and
- packet switched (PS).

A transmission channel can be allocated to be circuit or packet switched, depending on the traffic characteristics of the service classes using the transmission channel. So, if we consider a classic telephony service, for this service class, it is suitable to allocate the circuit switched (CS) channels, which remain allocated for a voice connection for its entire duration. The CS channels can also be allocated for various data connections in accordance with the per-burst reservation domain. In this case, the allocated channels are not released after the end of a connection, but they remain allocated until the end of a data burst. However, this is not an efficient reservation method because of the transmission gaps, but it is necessary to ensure the required QoS guarantees for specific services, as is also mentioned in Sec. 6.1.1. On the other hand, packet switched (PS) channels can be allocated for transmission of one data packet only. After the transmission is completed, the channel is free and can be used for a new transmission, either as a packet or a circuit switched channel.

Possible strategies for channel allocation are described in Sec. 5.4.3 and it is concluded that the best network efficiency can be achieved when the channels are allocated dynamically. Thus, in accordance with current needs of different subscribers in the network to apply various telecommunications services, they use transmission channels allocated for various service classes. However, demands of the network subscribers for using different service classes, as well as the traffic characteristics of the services used vary with time, possibly causing a frequent change of the channel allocation division. Accordingly, the network stations have to be frequently informed about a new channel order to be able to access the proper transmission channels allocated to a service class they use. For this purpose, the base station, which only has some knowledge of the channel order, has to inform network stations about an actual channel order by using a special signaling message.

6.1.3.2 Access to the Circuit Switched Channels

Network stations use the signaling channels to request different services. In the case of a service using CS channels (e.g. telephony), the allocation message (Sec. 6.1.2) sent by the base station contains the identification number of one or more transmission channels that are allocated to a particular station for the entire duration of the connection. After the connection is completed, the used channels are again free, as explained above.

In the case of disturbances in a CS channel, it is moved into error state (Fig. 6.3) and it has to be exchanged by another transmission channel. To inform the affected network station about the channel change, the base station has to send an additional reallocation message specifying the new transmission channel for the affected connection. A new channel is usually taken from the pool of reserved channels. However, a PS channel can also be reallocated to serve as a CS channel. So, it can be used for substitution if services using the circuit switched channels have a higher priority, such as in the example of a CAC strategy for PLC, presented in Sec. 5.4.3.

6.1.3.3 Access to the Packet Switched Channels

Access to the packet switched channels can be organized in the same way as for the circuit switched channels. However, in the case of PS channels, there could be a time period between the reception of a request from a network station and the beginning of the actual data transmission. This can happen because some data from other network stations, which has already completed the signaling procedure, have to be transmitted first and these transmissions are not yet finished.

One possibility is to inform the network station about its latest transmission rights before it can start transmitting (separated control messages, Fig. 6.2). However, as mentioned in Sec. 6.1.2, this approach causes a higher signaling load in the network and the probability that a signaling message is corrupted or it will get lost because of disturbances is higher. Additionally, owing to the dynamic change of the channel order in a network, the base station is not able to calculate an exact moment when transmissions will be completed and it is not possible to transmit the allocation message to a network station before the end of another transmission. This causes transmission gaps, in this case originating from the kind of signaling procedure.

Another possibility for the access control of the packet switched channels is the application of signaling procedure with joint control messages (Fig. 6.2) combined with a distributed access control mechanism. In this case, the waiting stations, that is, stations that have already received an allocation message together with an acknowledgment from the base station, observe the situation in the network and accordingly calculate themselves a new time for the beginning of a transmission. Figure 6.4 presents a distributed algorithm for the access control, proposed in [Hras03] for application in PLC networks.

It is assumed, that a user packet to be transmitted (e.g. IP packet) is segmented first into smaller data units (segments) that fit into so-called data slots, which are accessible sections of network resources, provided by a multiple access scheme as time slots, frequency bands or code sequences (Sec. 5.2). Thus, with a transmission request, a network station demands a number of the data slots in accordance with the size of a packet to be transmitted. The allocation message, sent by the base station and specifying the access rights, contains a number of data slots that have to be passed by the station before it starts to send (SP – slot to be passed, Fig. 6.4). The slots to be passed are used by other network stations that made the reservation earlier. The SP counter of a waiting station is decreased by 1 for every passed data slot belonging to a logical transmission channel that is allocated to a relevant service class. If the counter is zero, the station can start the transmission in the next available data slot that belongs to its service class.

Thanks to the distributed access control mechanism, a waiting station always stores information about the number of data segments that have to be transmitted by other stations before it starts to send, independent of the changing number of packet switched

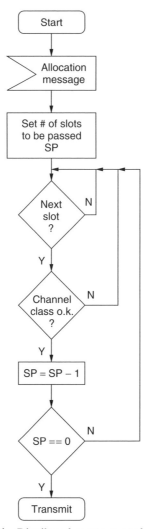

Figure 6.4 Distributed access control algorithm

channels. This also ensures that a waiting station starts the transmission immediately after a previous transmission is completed, which improves the network utilization as well. The same distributed access control algorithm can also be applied to every service. So, a number of separated algorithm instances can be used to serve multiple service classes.

However, application of a distributed access control mechanism is disadvantageous in networks in which the network stations can be selectively disturbed, as described in Sec. 6.1.2. In this case, several stations are affected by the disturbances and they are not able to recognize the channel order or, to count the data slots correctly if they are waiting stations. At the same time, other network stations that are not affected by the selective disturbances operate as usual. Thus, the disturbed networks can access the transmission medium at the wrong time, causing unwanted packet collisions. Therefore, the distributed access control mechanism has to be additionally protected against selective disturbances. For example, if a sending or waiting station is not able to recognize a transmission channel, and accordingly is not able to correctly count the passed data slots, it has to interrupt its transmission, or to interrupt execution of the access algorithm and to retransmit its request.

6.1.4 Signaling MAC Protocols

As mentioned above, a portion of the network resources needed for the realization of the reservation MAC protocols is allocated for the signaling procedure. For this purpose, one or more logical transmission channels are allocated for the signaling (Fig. 6.3). A signaling channel also presents a shared transmission medium, which is only used for the transmission of the control messages. The downlink of the signaling channel is fully controlled by the base station, as it is the case with data channels, and there is no need for access organization between multiple users. On the other hand, there is a need for access organization in the signaling uplink, which is specified by a MAC protocol. Any of the resource sharing strategies analyzed in Sec. 5.3 can be applied as a signaling MAC protocol as well.

Reservation MAC protocols are widely used in the existing cellular mobile communications systems (e.g. GSM, GPRS) and they are also proposed to be applied in the next generation of wireless networks (e.g. WATM, WLAN, UMTS). Accordingly, there is a lot of research work on this topic, as well as many standards and implementation solutions proposed for various network realizations. The main difference between various protocol solutions, is the application of different MAC protocols to the signaling channel. This section summarizes different approaches for the realization of the signaling MAC protocols.

The most applied reservation method is the ALOHA-based reservation procedure because of its simplicity. PRMA (Packet Reservation Multiple Access) and DPRMA (Dynamic PRMA) protocol have been developed for WATM (Wireless ATM) networks and uses the slotted ALOHA method for transfer of the transmission requests from mobile stations to the base station [Pris96], [AkyiMc99]. Very often, so-called minislots are used for contention based transmission of the requests (e.g. Centralized PRMA – CPRMA, [AkyiMc99]). It improves performance of the reservation procedure because of the usage of a smaller portion of network capacity for signaling.

The signaling procedure can also be organized in a dedicated manner. In this case, network stations can only use fixed predefined request slots to send their transmission demands to the base station. The dedicated request procedure is usually organized according to the polling access method, as described in [AcamKr]. The base station sends the polling messages to the network stations, for example, according to the round-robin procedure, and a station can then send a transmission request only after it has received a polling message. Polling-based reservation protocols are considered for use in satellite networks [Peyr99]; for example, Priority-Oriented Demand Assignment (PODA) and Mini-Slotted Alternating Priorities (MSAP) protocols.

Both contention- and arbitration-based reservation protocols can be extended to provide better network performance. An often applied protocol extension is piggybacking (see e.g. [AkyiMc99] and [AkyiLe99]). In this case, a transmission request can be added to the last data segment of a currently transmitted packet (piggybacked). So, the current packet transmission is also used for a contention-free request of the next packet and no additional network resources are used for the reservation.

The application of ALOHA-based reservation methods has an advantage because of the protocol simplicity. However, instability of such a protocol increases with the network load, as shown in Sec. 5.3.2. On the other hand, polling-based protocols behave better if the network is highly loaded, but cause long round-trip times of the polling messages if the number of network stations is high. In order to merge the advantages of both ALOHA and polling-based protocols, and also to improve the protocols by avoiding their disadvantages, hybrid access protocols, which use both the random and the dedicated access method have been developed, as mentioned in Sec. 5.3.3 as well. An example is Identifier Splitting Algorithm combined with Polling (ISAP – protocol), which uses a collision resolution method for the reservations [HoudtBl00]. After a certain level of the resolution tree is reached as a result of frequent transmission requests and collisions, the protocol switches to the polling-access method. There are also protocols that switch from the random-access method to the reservation method. So, in the Random Access Demand Assignment Multiple Access (RA/DAMA) reservation protocol, the remaining network resources can be used with random access [ConnRyu99]. Another example is Distributed Queuing Random Access Protocol (DQRAP), which changes to reservation after an unsuccessful random request [AlonAg00].

A further group of reservation protocols can be described as adaptive protocols, which change the access method according to the current network status (network load). A typical adaptive reservation protocol is Minimum-Delay Multi-Access protocol (MDMA) [Peyr99]. According to the current network load, the MDMA protocol calculates the probability, determining if a transmission will be carried out with the random access or the reservation method. In some cases, the reservation and the random-access methods are carried out simultaneously. In [Bing00], it is shown how an ALOHA-based protocol can be stabilized using a variable number of request slots, depending on the network load. A further possibility for stabilization of ALOHA protocols is the increase in the retransmission time after collisions [DengCh00], as described in Sec. 5.3.2.

An overview of the reservation MAC protocols, presented above, shows that there are many possible protocol solutions and their derivations, which could also be applied to the PLC access networks. Generally, the two basic methods for the signaling MAC protocol are realized with the following principles:

- random access – realized by slotted ALOHA, and
- dedicated access – realized by polling.

Both ALOHA and polling-based reservation protocols have some disadvantages, if they are applied in their basic forms. Therefore, the basic protocol solutions are very often extended to improve their performances. The protocol extensions can be classified as follows:

- Piggybacking,
- Hybrid protocols – a combination between both basic protocol solutions, or a combination between the random access method and the reservation, and
- Adaptive protocols – access parameters change according to the current network load/ status (number of accessible request units, variation of mean retransmission time, and so on).

As presented above, there are numerous proposals and standards considering reservation MAC protocols. However, they are mostly developed for wireless communications systems and are adapted to the existing or proposed wireless standards. Therefore, there is a need for an investigation of reservation MAC protocols for their application in PLC access networks, which considers particular characteristics of the powerline transmission medium and its environment (presented in Chapter 3), as well as features of the PLC transmission systems (Sec. 4.2). In particular, the robustness of the reservation protocols to disturbances has to be investigated, as well as possible solutions for realization of the tasks for fault management, which have to be integrated within the protocols. On the other hand, the protocol extensions that improve network performance have to be analyzed, to consider their true advantages in a communications system such as PLC. Implementation complexity, and application under unfavorable disturbance conditions have to be analyzed as well. Finally, different protocol solutions have to be compared fairly and under the conditions that are expected in the PLC access network. A detailed performance analysis of reservation MAC protocols under specific PLC conditions is presented below.

6.2 Modeling PLC MAC Layer

In Sec. 6.1.4, we presented several options for access organization in the uplink part of the signaling channel and concluded that it is necessary to investigate various solutions for the signaling MAC protocols to be applied to the PLC access network. In this section, we present our approach to model the PLC MAC layer and to carry out a performance evaluation of the signaling MAC protocols. A MAC protocol applied to a network with multimedia traffic, such as PLC, has to ensure sufficient QoS for different kinds of telecommunications services and for good network utilization. Therefore, we compare the performance of various protocol solutions in accordance with QoS level, which can be provided by their application.

Relevant QoS parameters to be observed in the performance analysis and different modeling approaches are discussed in Sec. 6.2.1. Applied simulation model, representing a PLC MAC layer, including disturbance and user models, as well as some protocol assumptions defined for this investigation, are presented in Sec. 6.2.2. Finally, in Sec. 6.2.3, we define traffic models that are used in the investigation and present the simulation technique and simulation scenario applied in this performance analysis.

6.2.1 Analysis Method

6.2.1.1 Investigation of Relevant QoS Parameters

As mentioned in Sec. 5.4.3, the QoS requirements for various services can be specified by several performance parameters. Network providers usually guarantee the following QoS parameters, which are relevant to the subscribers and their judgment of the network quality: blocking probability, data rate or data throughput, loss probability, transmission delays on different network levels, and dropping probability.

A reservation MAC protocol manages the signaling procedure, the transfer of the transmission requests and acknowledgments. Once the reservation procedure is completed, the transmission takes place independently of the signaling procedure. Therefore, dropping probability as well as loss probability do not characterize the performance of the reservation protocols, and both events take place during transmission after the reservation procedure is completed and they can only be controlled by a mechanism for traffic scheduling (Sec. 5.4.2). Similarly, the blocking probability can be controlled by a CAC mechanism, as shown in Sec. 5.4.3. On the other hand, the blocking of a connection can also be caused if the signaling procedure is not successful or if it is too long. So, there is a direct influence of applied reservation protocol on the blocking probability. However, the blocking events can be observed through the delays caused by the signaling procedure.

The fulfillment of the QoS parameters is important for ensuring the subscriber requirements. On the other hand, network providers try to use available network efficiently to be able to serve a larger number of subscribers. Accordingly, an important provider relevant parameter is network utilization, which is especially related to the networks with limited data rates, such as PLC access networks. Therefore, network utilization is also one of the parameters that characterizes the performance of the reservation MAC protocols. So, we can conclude that a performance analysis of reservation MAC protocols can be carried out by observation of the following three QoS parameters:

- network utilization,
- transmission delay, and
- data throughput.

6.2.1.2 Modeling Technique

The performance analysis of a communications network and the evaluation of QoS parameters, such as the relevant QoS parameter defined for this investigation, can be carried out by the following three methods [Fort91]:

- measurements,
- analytical modeling, and
- simulation modeling.

Simulation modeling is chosen as the primary analysis method in this investigation for the following reasons [Hras03]:

- complexity of the reservation MAC protocols,
- variety of applied disturbance and traffic models,
- fair performance comparison of different protocol solutions, and
- the possibility of a detailed investigation of the protocol implementation.

Reservation MAC protocols seem to be complex and they usually consist of several protocol components, as described in Sec. 6.1. Accordingly, it is very difficult to represent such complex protocols in the analytical models and various assumptions has to be done. On the other hand, by using simulation modeling it is possible to achieve a needed representation grade. The modifications of individual protocol components can cause significant changes of performances and behavior of particular QoS parameters. Therefore, there is the need for the optimization of different protocol components to realize an efficient protocol solution. Such tuning of numerous parameters of a complex protocol can be carried out more efficiently by simulation modeling rather than by analytical methods.

An investigation of MAC protocol under multimedia traffic calls for performance analysis using various service and traffic models (Sec. 4.4.2). The traffic models can be very simply implemented within a simulation model, as well as various combinations of source models representing different telecommunications services. Traffic traces, achieved by measurements in real communications networks, can also be used as source models in simulations. In the same way, simulation methods are also suitable for disturbance modeling (Sec. 3.4.4). So, within a simulation model, it is possible to implement multiple error models representing different kinds of disturbances and caused by various noise sources.

The investigation of different protocol solutions has to ensure their fair comparison under the same modeling conditions. At the same time, the investigated protocol solutions are complex and any approximation of protocol functions, which is necessary to simplify the analytical modeling, can vary the achieved results. Therefore, the investigated protocols have to be implemented in detail, which can be made easier in a simulation model [Müller02].

Currently, there are no standards that specify a transmission system to be applied to broadband PLC access networks (Sec. 2.2.3). Therefore, the simulation model defined in Sec. 6.2.2 represents a theoretical proposal for the PLC MAC layer, which is used as a logical model for the investigation of signaling MAC protocols. Accordingly, several parameters characterizing this access scheme are assumed; for example, data rate of the transmission channels, duration of a time slot, segment structure, and so on. However, a simulation model made for investigation protocols on the logical level can be easily adapted to any parameter set and access scheme, allowing investigation of the MAC layer under specific system conditions. Finally, a further advantage of simulation modeling is a simultaneous consideration of the implementation complexity and the reliability of the investigated protocols, because of the need for their implementation within the simulation model (software integration). However, simulation models to be implemented for this purpose are also complex and therefore they have to be carefully validated.

6.2.1.3 Analytical Modeling and Empirical Performance Analysis – Examples

An analytical approach for modeling of a PLC network at the call/burst level is proposed in [BegaEr00]. The analytical model corresponding to the PLC network model, presented in Sec. 5.4.3, including a proposed CAC mechanism and its admission policy is implemented by using a standard tool for performance modeling MOSEL [BegaBo00]. Disturbances are modeled to occur independently on different logical transmission channels in accordance with the on–off disturbance model, presented in Sec. 3.4.4. Depending on the offered traffic load in the PLC network, caused by two kinds of services – telephony and Internet data transfer – as well on as the available data rate in the networks affected by the disturbances, the following QoS parameters are observed:

- blocking probability of voice calls,
- average data rate of Internet connections,
- network utilization, and
- channel availability; that is, how many logical transmission channels are affected, or not affected by the disturbances.

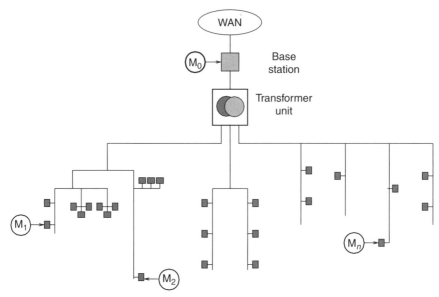

Figure 6.5 Performance measurements on a real PLC network for empirical analysis

The empirical investigations of MAC protocols can be carried out on test equipment that provides the possibility of implementing various protocol solutions and measurements needed for the observation of QoS parameters, or on a real PLC access network. A performance analysis on the real system can be carried out on the MAC layer, as well as on any other network layer, depending on particular aims of an empirical study. Network utilization and data throughput, as relevant QoS parameters for performance investigation of the reservation MAC protocols, are measured in the base station at the interface between a PLC network and its distribution network, used for connecting with the WAN (e.g. measurement point M_0, Fig. 6.5). In this way, it is possible to consider the PLC part of the common communications structure. Thus, eventual bottlenecks and other effects in the distribution network cannot influence results of the measurements.

For the measurements of various packet delays in a PLC network, it is necessary to establish one or more additional measurement points within the PLC network ($M_1 - M_n$). Thus, it is possible to estimate the delays between the base station and each network user in both transmission directions. However, to ensure an exact delay estimation, the measurement equipment has to be strictly synchronized. The placement of the measurement points within a PLC network has to be done to ensure an observation of specific effects in the network. So, the measurement equipment can be placed on the premises of subscribers at different distances from the base station, or in various network segments that are connected to the base station over different numbers of repeaters, and so on.

6.2.2 Simulation Model for PLC MAC Layer

6.2.2.1 Generic Simulation Model

The simulation model, developed for the investigation of signaling MAC protocols [Hras03], [HrasHa00], [HrasLe00a], represents an OFDMA/TDMA scheme (Sec. 5.2.2, Fig. 5.8).

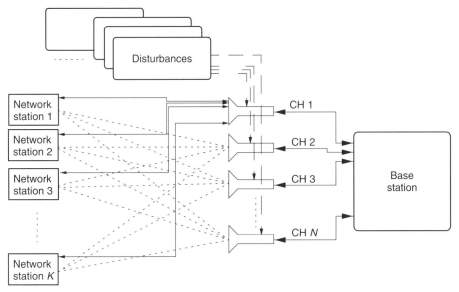

Figure 6.6 Generic simulation model

There are a number of bidirectional transmission channels that connect network users/subscribers with the base station (Fig. 6.6), which lead to the FDD mode, with symmetric division of data rates between uplink and downlink transmission directions. As mentioned in Sec. 5.4.1, duplex modes with asymmetric and dynamic division provide better network performance than the fixed mode. However, this investigation considers the MAC protocol for the signaling channel, which can be analyzed independently of the applied duplex mode and division strategy. The results achieved in the simulation model with FDD are valid for networks with TDD mode as well.

The transmission channels can be accessed by all network stations in the uplink transmission direction (shared medium) while the downlink is controlled by the base station. There is a possibility for the modeling of various disturbance types, which can be implemented to affect both single and multiple transmission channels (Sec. 3.4.4) and to represent different types of noise. The subscribers are represented by the network stations that provide multiple telecommunications services (e.g. telephony and Internet). Network stations and base stations implement all features of the investigated MAC layer and protocols, including multiple access scheme, MAC protocol for the signaling channels, the signaling procedure and the access control, mechanisms for error handling, and so on.

The generic simulation model is designed to represent the OFDMA/TDMA scheme. However, the model can be easily adapted to represent a TDMA system as well as any combination of TDMA and FDMA methods. On the other hand, the evaluated performance of the signaling MAC protocols can be interpreted independently of the modeled multiple access scheme, by a generalization of the simulation results. An implemented shared communications medium (PLC medium) can also be used for the modeling of other networks with similar communications organization and equivalent transmission features (e.g. mobile wireless networks).

6.2.2.2 Disturbance Modeling

As presented in Sec. 3.4.4, the disturbances in PLC networks can be represented by an on–off model:

- OFF – the channel is disturbed and no transmission is possible, and
- ON – the channel is available.

These two states are modeled by two random variables that represent interarrival times and durations of the disturbances. Both random variables are assumed to be negative exponentially distributed. The following three disturbance scenarios are used in further investigations [Hras03], [HrasHa01]:

- disturbance-free network,
- lightly disturbed network – 200 ms mean interarrival time of the impulses/disturbances, and
- heavily disturbed network – 40 ms mean interarrival time of the disturbances.

The mean duration of a disturbance impulse is set to $100\mu s$ and it is assumed that the noise impulses with a duration shorter than $300\mu s$ do not cause transmission errors (e.g. owing to symbol duration, FEC, Sec. 4.3). In this investigation, the disturbances are modeled independently for each transmission channel (Fig. 6.6).

6.2.2.3 User Modeling

To be able to model various services, network stations implemented in the simulation model (Fig. 6.6) can be connected with a number of traffic models, to represent different telecommunications services or various service classes (Sec. 4.4). Both primary telecommunications services, Internet-based data transmission and telephony, representing a packet switched and a circuit switched service respectively, are implemented in the simulation model, as shown in Fig. 6.7. The packets (e.g. IP packets) from the data traffic source are delivered to the packet queue of the network station. Later, the packet is segmented into segments, which are stored in the transmission queue. Both packet and transmission queues can store exactly one packet (or segments of a packet). So, a maximum of two user packets can be stored in the network station. Note, the reservation is always carried out for one user packet in accordance with the per-packet reservation principle (Sec. 6.1.1). After successful transmission of the packet, the next packet (if any) is moved from the packet to the transmission queue. Later, the reservation procedure is carried out for the new packet.

The data source generates user packets according to an applied traffic model. After each packet generation, the data source calculates a time for the generation of a new packet. If the packet queue is occupied, the data source is stopped and it can deliver the new packet after the packet queue is empty again. In this case, the following two situations are possible:

- The packet queue is emptied before the new packet has to be generated and the new packet can be delivered at the calculated time.

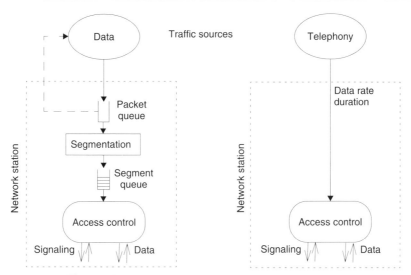

Figure 6.7 User models for data and telephony service

- The data source is still stopped at the packet generation time. In this case, the new packet is generated immediately after the packet queue is emptied.

The implementation of the telephony service is simpler (Fig. 6.7). The calls are generated in accordance with a traffic model for the telephony; for example, as specified in Sec. 4.4.2. Generally, for a circuit switched service (CS), the necessary data rate (if different CS services are modeled) and duration of a connection have to be calculated by the traffic model. The reservation procedure is the same as in the case of the data service. However, the signaling procedure is carried out only once in the case of a CS connection for its setup.

6.2.3 Traffic Modeling

6.2.3.1 Modeling Telephony Traffic

As mentioned above, the signaling procedure is carried out for each transmitted packet, in the case of data connections, in accordance with the per-packet reservation principle, and once per connection for the telephony, according to the per-connection reservation principle. In Sec. 4.4.2, it is shown that the arrivals of the voice connections seem to be in a range of minutes, whereas the data packets, for example, caused by an Internet connection, are generated in a range of seconds. Thus, it can be recognized that the arrival rate of the voice connections is significantly lower when compared with arrivals of IP packets. Accordingly, the arrival rate of the transmission requests for data transmissions is much higher than in the case of telephony. Therefore, the requests for telephony connections can be omitted in this investigation and they are not particularly modeled.

6.2.3.2 Simple Internet Traffic Models

Data traffic is characterized by two random variables; mean packet size and interarrival time of packets. To define a simple model for Internet-based data traffic, the mean packet size is set to 1500 bytes in accordance with the maximum size of an Ethernet packet. The mean interarrival time of packets represents user requests for download of WWW pages and it is chosen to be 4.8 s. So, the average data rate per subscriber amounts to a relatively low value of 2.5 kbps. However, in other studies considering Internet-based data transmission in the uplink direction, the offered network load per user is even lower (e.g. 662.5 bps [TrabCh]).

The data packets transmitted in the uplink are usually small IP packets representing control and request packets with an average size between 92.9 and 360 bytes respectively [TrabCh]. Therefore, there is a need for a second simple traffic model with shorter data packets. The mean packet size for the second model is set to 300 bytes. To be able to compare networks with small and large packets, both source models have to produce the same average offered network load per subscriber (Tab. 6.1). Therefore, the mean interarrival time of packets for the second model has to be 0.96 s. This interarrival time represents the time between user requests for the downloads and seems to be too short. However, during a download there are a lot of automatically produced requests for so-called in-line objects that are contained in a WWW page (Sec. 4.4.2). Additionally, there are a large number of control packets, caused by the acknowledgments provided by TCP.

The arrival of the data packets is very often described as a Poisson process and negative exponential distributions are usually used for modeling the interarrival time (e.g. [AlonAg00], [FrigLe01a]). Because of the applied time-discrete simulation tool (Sec. 6.2.4) in this investigation, the interarrival time in the simple traffic models is modeled as a geometrically distributed random variable. The packet size is modeled as a geometrically distributed random variable as well. The application of two simple traffic models with different interarrival times of packets offers the possibility of investigating the reservation MAC protocols under rare and frequent transmission requests. Because of the chosen per-packet reservation principle, the protocol performance is expected to vary, depending on the applied traffic model. In the case of frequent requests, the signaling channel has to transmit five times more requests than in the case of rare requests. Accordingly, the signaling channel is significantly more loaded than in the case of rare transmission requests.

The simple traffic models provide the variable packet sizes, which are geometrically distributed. However, in the real world, computers and other communications devices for data transfer operate with discrete sizes of the packets (e.g. IP packets). In general, there are only a few possible packet sizes between a minimum and a maximum value. Therefore, the modeling of realistic or nearly realistic Internet traffic on the IP level has to be carried out by the application of the so-called multimodal traffic models described

Table 6.1 Parameters of simple traffic models

Parameter/model	Rare requests	Frequent requests
Mean packet size	1500 bytes	300 bytes
Mean interarrival time	4.8 s	0.96 s
Offered network load per user	2.5 kbps	
Packet size – distribution	Geometric	
Interarrival time – distribution	Geometric	

below. However, in [HrasLe03], it is shown that the simple traffic models represent a very good approximation of the user behavior compared with the multimodal models. On the other hand, negative exponential distributions applied in the simple traffic models are convenient for the analytical modeling approaches.

6.2.3.3 Multimodal Traffic Models

As mentioned above, a big part of the traffic load in the uplink belongs to the WWW requests. The size of the IP packets carrying WWW requests is differently specified in various traffic models; for example, between 64 bytes in [HoudtBl00] and 344 bytes in [TrabCh]. The share of manually generated requests swings between 10 and 50% of all packets in the uplink, and the part of the automatic request is between 38 and 88% [Arli]. Other types of packets transmitted in the uplink as well as in the downlink are the control packets. The size of the control packets, which are mainly caused by the transmission of acknowledgments used in the TCP protocol, is considered to be between 40 and 92 bytes [TrabCh].

For the multimodal traffic model to be used in the investigation of the MAC layer in the uplink transmission direction in the PLC access networks [Hras03], [HrasLe03], it is assumed that 85% of the packets in the uplink are control and request IP packets (Fig. 6.8). The remaining 15% are larger IP packets; for example, caused by the transmission of chart messages, e-mail transfer, and so on. The small packets can have two sizes, 64 and 256 bytes, which are generated with the probabilities 0.45 and 0.40 respectively. The maximum size of the IP packets specified in Ethernet LAN networks is about 1500 bytes, and their probability is set at 0.1. The probability of 1024 byte packets, representing other transmissions of the larger packets, is 0.05. All larger files to be transmitted over the network are segmented into multiple IP packets.

The mean size of the packets generated according to the uplink traffic model is 332.5 bytes. The interarrival time is a geometrically distributed random variable according to the negative exponential distribution proposed in most multimodal IP traffic models (e.g. [ReyesGo99]).

The subscribers of PLC access networks can also offer some Internet content (various information, publications, music or video files, etc.) that are downloaded by users

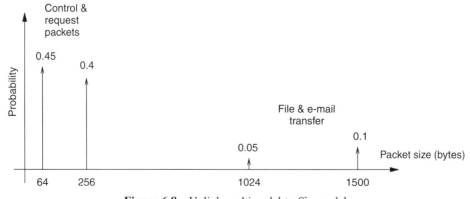

Figure 6.8 Uplink multimodal traffic model

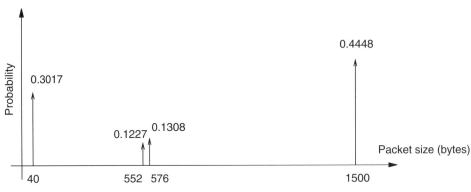

Figure 6.9 Downlink multimodal traffic model

placed out of the PLC network. In this case, the traffic characteristic of such PLC subscribers is different from the typical uplink characteristics and it can be represented by the source models characterizing typical Internet traffic in the downlink. For this purpose, we can use a multimodal traffic model proposed in [ReyesGo99] for investigation of Internet traffic for wireless systems in the downlink transmission direction (Fig. 6.9). According to the characteristics of the downlink Internet traffic, the most frequent packets are 1500 bytes IP packets (44.48%). This is caused by downloads of larger files/pages, which are segmented into the Ethernet packets with the maximum size. The second more frequent type are 40-byte control packets (30.17%). They are empty packets caused by TCP/IP protocol and include only the overhead information (e.g. acknowledgments).

The mean packet size is 822.33 bytes and interarrival time is also a negative exponentially distributed random variable.

6.2.3.4 Modeling Various Data Services

The multimodal traffic models characterize the IP traffic caused by WWW applications. However, an ordinary subscriber also uses other data services and applications. A very frequent data application is electronic post service (e-mail). The e-mail messages present usually larger data files with an average size of several hundreds of bytes [TranSt01]. However, the frequency of the e-mail messages transmitted in the uplink is significantly lower than is the case with the WWW requests. Accordingly, intensity of the transmission requests caused by e-mail is low as well. So, it can be assumed that e-mail traffic is also represented in the multimodal traffic model for the uplink, by large packets (e.g. 1500 and 1024 bytes) with relatively low generation probability.

A further very common data application is FTP (File Transfer Protocol), used for downloads of different files from remote servers. However, the usage of the FTP decreases with growing WWW traffic, which also provides the same download functions [TranSt01].

The transmission of video traces and files is a popular application, which is expected to increase rapidly in the near future. The data rates caused by video are much higher than those for the WWW traffic (mean of 239 kbps [AkyiLe99]) and the traffic characteristic is represented by a nearly continuous data transmission, which corresponds to the typical behavior of the streaming traffic class. The streaming services, such as video and audio transmission, cause a higher network load in the downlink transmission direction. In the uplink, the control messages are transmitted with an intensity that depends on the variety of the streaming data rates [FrigLe01a]. However, the control messages are represented with a high generation probability in the multimodal WWW traffic model for the uplink (Fig. 6.8) as well. So, this can be used as an approximation for the modeling of the streaming uplink. On the other hand, the uplink multimodal model (Fig. 6.9) with a higher probability for large packets can be applied as an approximation for the streaming downlink, in the case in which a streaming server is situated in the PLC access network.

Various Internet games belong to a further growing group of communications services (e.g. [Bore00]). In this case, the playing subscribers are involved in a permanent Internet connection with the game server and/or a number of other players. The number of available Internet games is rapidly increasing and it is very difficult to specify traffic models that can represent this part of the Internet traffic. However, the games consist mainly of a very intensive exchange of short packets/requests between involved subscribers, which could also be approximated by high generation probabilities of short packets in the multimodal traffic models.

6.2.4 Simulation Technique

6.2.4.1 Implementation of the Simulation Model

The simulation model used in this investigation is implemented using YATS (Yet Another Tiny Simulator [Baum03]), a tool developed at the Chair for Telecommunications, Dresden University of Technology. YATS is a discrete-time and discrete-event simulator tailored for various communications networks. It provides a number of modules that are used for investigations of ATM, DQDB, PLC and various wireless networks, as well as TCP/IP-based data traffic.

The YATS simulator provides several possibilities for validation of implemented network models and protocols. So, data objects can be traced through a network model and the change in their parameters can be observed. There are also possibilities for graphical presentation of interesting protocol parameters, which ensure operation test of the implemented protocols. YATS is also used as a basic platform for the PAN-SIM (PLC Access Network Simulator) tool, which is provided for performance evaluation of PLC access networks. A brief description of the PAN-SIM is presented below. A detailed description of the PAN-SIM can be found in [palas01a].

6.2.4.2 PLC Access Network Simulator

The PLC Access Network Simulator (PAN-SIM) was developed during the PALAS (Powerline as an Alternative Local AccesS) project, supported by the European Union. The main goals of the PAN-SIM are

- demonstration of PLC system behavior,
- study of the MAC layer for the uplink in a PLC network,
- study of channel disturbances and error-control mechanisms,
- performance evaluation of PLC systems under multimedia traffic, and
- planning of PLC access networks.

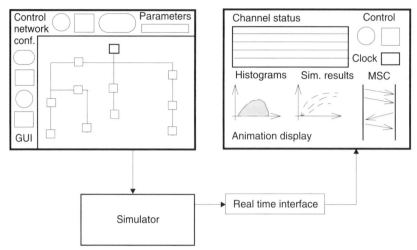

Figure 6.10 PLC access network simulator

There are three main parts in the simulation tool: simulator (PAN-SIM kernel), Animation, and Graphical User Interface (GUI), as shown in Fig. 6.10. The simulator kernel includes the implementation of the PLC transmission system, disturbance scenarios, MAC layer and access protocols, as well as the error-handling methods. The simulator is connected by a real-time interface to the animation part of the PAN-SIM. It allows presentation of the simulation results, as well as the possibility of observing interesting protocol parameters and network behavior on the animation display.

The animation presents the operation of the implemented protocols, which are usually very complex, and in this way makes their testing easier. Observation of the protocol and network behavior gives an additional possibility for their analysis as well. Animation display can present histograms, different simulation results, flow diagrams of implemented protocols, current channel status, and so on. Finally, a goal of the PAN-SIM is its usage for the presentation purposes of the PLC access network, which is also realized by animation. PAN-SIM has been presented at several trade fairs and conferences; for example, CeBIT2001, ISPLC2001, CeBIT- Asia2001.

The graphical user interface serves as a user friendly control platform for the PAN-SIM and as a tool for network configuration. A graphical editor allows very fast and easy network configuration with all its elements and parameters. Once a PLC network is configured or loaded from a configuration file, it can be investigated with the use of the simulator. Furthermore, the GUI ensures a convenient way to set, change and modify various parameters (mentioned below), which are necessary for the simulation:

- simulation parameters,
- parameters describing network structure and topology,
- protocol parameters,
- parameters needed for traffic models, and
- disturbance parameters.

The network structures defined by the GUI can also be used as an input for other tools, for example, analytical tools for performance analysis or network planning tools.

6.2.4.3 Simulation Scenario

A performance evaluation of various solutions for the signaling MAC protocols has to be carried out in network models with varying traffic conditions. Thus, it is possible to investigate features of the MAC protocols under different network load conditions. To vary the network load, the number of network stations is increased from 50 to 500. This results in a minimum average network load of 125 kbps and a maximum of 1.25 Mbps, in accordance with the simple traffic models presented in Sec. 6.2.3. Another approach to the increase of the network load is a variation of offered traffic for individual network stations; for example, the offered network load of individual network stations can be varied from 2.5 to 25 kbps for a constant number of stations, which results in the same common offered network load, as in the first case.

If the number of stations remains constant, the interarrival times of the user packets has to be reduced to increase the network load. That means, for a network load of 1.25 Mbps and 50 network stations, the interarrival time has to be set to 480 ms in the simple traffic model with rare requests and to 96 ms in the model with frequent requests. So, the interarrival times would become too short and the representation of a realistic WWW traffic scenario disappears. On the other hand, the average intensity of the transmission requests is equal in both cases – a variable and a fixed number of the network stations – if the common network load remains the same.

A transmission request is made only after a previous packet transmission is successfully completed (Sec. 6.2.2). On the other hand, if the number of network stations is increased, the number of uncorrelated sources in the network becomes higher. Accordingly, the common number of transmission requests is higher, which is not the case if the number of network stations is constant. Therefore, the increasing number of network stations also presents a worse case for the consideration of the reservation MAC protocols with per-packet reservation domain and is chosen to be used in further investigations.

6.2.4.4 Parameters of the Simulation Model

In Sec. 6.2.3, it is concluded that the consideration of the telephony service is not relevant to the investigation of the reservation MAC protocols and the requesting procedure for telephony does not have to be modeled. The classical telephony service uses circuit switched transmission channels provided by the OFDMA scheme. For this investigation, it is assumed that one half of the network capacity is occupied by telephony and other services using the circuit switched channels. The remaining network capacity is occupied by the services using packet switched transmission channels.

Recent PLC access networks provide data rates of about 2 Mbps. If the data rate of a transmission channel is set to 64 kbps, there will be approximately 30 channels in the system. Accordingly, the number of packet switched channels in the model is 15, which results in 960 kbps net data rate in the network (Tab. 6.2). One of the transmission channels is allocated for signaling, which is necessary for the realization of the reservation procedure. The duration of a time slot provided by the OFDMA/TDMA (Sec. 5.2.2) scheme is set to 4 ms in the simulation model. Within 4 ms, a 64-kbps transmission channel can transmit a data unit of 32 bytes. Accordingly, the size of a data segment is also set to 32 bytes. It is also assumed that the segment header consumes 4 bytes of each segment, so that the segment payload amounts to 28 bytes.

Table 6.2 Parameters of the simulation model

Parameter	Value
Number of channels	15
Number of signaling channels	1
Channel data rate	64 kbps
Time-slot duration	4 ms
Segment size	32 bytes = 4 bytes header + 28 bytes payload

The duration of a simulation run is chosen to correspond to the time needed for at least 10,000 events (generated packets) in the network. Also, 10 simulation runs and a warm-up run are carried out for each simulation point – the network load point is determined by the number of stations (e.g. between 50 and 500). From the simulation runs, the mean value, the upper bound, as well as the lower bound of the 95% confidence interval, are computed and included in all diagrams representing the simulation results.

6.3 Investigation of Signaling MAC Protocols

An overview of the existing reservation MAC protocols, given in Sec. 6.1.4, shows that there are many protocol solutions and their derivatives that are investigated for implementation in different communications technologies. However, according to the chosen resource sharing strategy (MAC protocol) to be applied to the signaling channel, two protocol solutions can be outlined as basic reservation protocols:

- protocols using random access to the signaling channel, mainly realized by slotted ALOHA, and
- protocols with dedicated access, usually realized by polling.

Performance analysis of the basic protocols presented in Sec. 6.3.1 is carried out with the following two aims: investigation of the basic protocols in a PLC transmission system specified by its multiple access scheme (in this case OFDMA/TDMA) in a typical PLC environment, characterized by unfavorable disturbance conditions, and validation of used simulation model and chosen investigation procedure. Further, in Sec. 6.3.2, we analyze several protocol extensions, and finally in Sec. 6.3.3, we present a performance analysis of advanced polling-based reservation MAC protocols, which are outlined to achieve the best performance in the case of per-packet reservation domain.

6.3.1 Basic Protocols

6.3.1.1 Description of Basic Reservation MAC Protocols

The transmission channels provided by the OFDMA/TDMA scheme are divided into time slots that can carry exactly one data segment (Sec. 5.2.2). It is also the case in the signaling channel, which is divided into request slots in its uplink part and control slots in the downlink. The request slots are used for transfer of the transmission request from the

Performance Evaluation of Reservation MAC Protocols

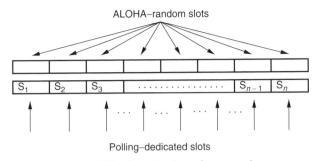

Figure 6.11 Organization of request slots

network stations to the base station, whereas the control slots are used by the base station for transmission of acknowledgments and transmission rights, as well as other control messages, as described in Sec. 6.1.2. In the case of ALOHA reservation MAC protocol, the request slots are used randomly (Fig. 6.11). On the other hand, the polling protocol uses dedicated request slots, which are allocated for each network station.

According to the ALOHA protocol, a network station tries to send a transmission request, containing the number of data segments to be transmitted to the base station, using a random request slot. In the case of collision with the requests from other network stations, the stations involved will try to retransmit their transmission demands after a random time (Fig. 6.12). After a successful request, the base station answers with the number of data slots to be passed before the station can start to send. According to

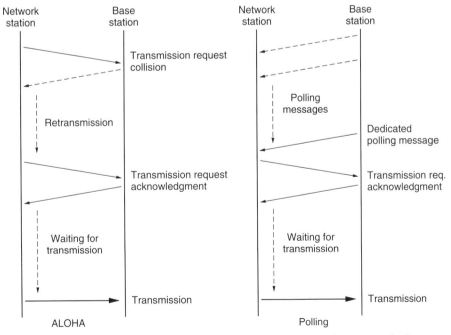

Figure 6.12 Order of events in ALOHA and polling-based access methods

the distributed allocation algorithm (Sec. 6.1.3), the station counts data slots to calculate the start of the transmission. The polling procedure is realized by the base station that sends so-called polling messages to each network station ($S_1 - S_n$) in accordance with the round-robin procedure. Only the station receiving a polling message has the right to send a transmission request. After a successful request, the rest of the signaling procedure is carried out, such as in the case of ALOHA protocol, by using the distributed allocation algorithm. The collisions are not possible, but a request can be disturbed and in this case, it has to be retransmitted in the next dedicated request slot.

In the case of ALOHA protocol, it is possible to transmit exactly one transmission request during a time slot. The acknowledgment from the base station is sent in the next time slot, if there is no collision (Fig. 6.13). According to the polling protocol, the base station can poll exactly one network station during a time slot, which also allows a request per time slot. Acknowledgment is transmitted in the next time slot after the request, such as in the ALOHA protocol.

Both ALOHA and polling protocols have the same procedural rules and a fair comparison can be made. Therefore, the base station has to be able to poll a network station and to send an acknowledgment during the same time slot. A polling message in slot i addresses a network station to send a transmission request in slot $i + 1$. At the same time, an acknowledgment in slot i confirms a request from slot $i - 1$.

6.3.1.2 Network Utilization

Network utilization is observed as a ratio between used network capacity for the data transmission and the common capacity of the PLC network. Only error-free segments are taken into account for used network capacity. In this part of the investigation, a simple packet retransmission method is implemented, in accordance with the send-and-wait ARQ mechanism (Sec. 4.3.4). So, in the case of an erroneous data segment, all segments of a user packet have to be retransmitted. Of course, the data segments that had to be retransmitted are not counted as used network capacity. The simple packet

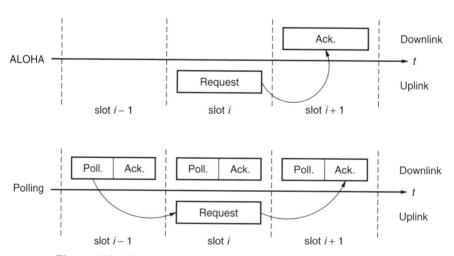

Figure 6.13 Slot structure for ALOHA and polling-based protocols

retransmission is not an efficient method for error handling. However, this approach ensures an observation of the protocol performance without influence of an applied error-handling method. Application of other ARQ variants that can improve network utilization are considered in Sec. 6.4.

If the networks with rare transmission requests are analyzed (average packet size of 1500 bytes in the simple data traffic model, Sec. 6.2.3), there is no difference between ALOHA and the polling reservation MAC protocols (Fig. 6.14). There is a linear increase in the network utilization from 15% to the maximum values. The maximum network utilization is reached within the network without disturbances (about 93%). The remaining 7% of the network capacity is allocated for the signaling channel (one of 15 channels). In the lightly disturbed network, the maximum utilization amounts to 83%, and in the heavily disturbed network, it is about 50%.

A saturation point can be recognized in the diagram between 300 and 350 stations in the network without disturbances. Each network station produces on average 2.5 kbps of offered traffic load (Sec. 6.2.3), which amounts to 750 to 875 kbps for 300 to 350 stations, according to Eq. 6.1.

$$L = n_{NS} \cdot l \qquad (6.1)$$

L – average total offered network load
n_{NS} – number of network stations
l – average offered load per station 2.5 kbps

The network has a gross data rate of 896 kbps (14 channels with 64 kbps). However, according to the size of the data segment payload (28 bytes, Sec. 6.2.4) and Eq. 6.2, it results in a net capacity of 784 kbps (14 channels with 56 kbps), which also has a total

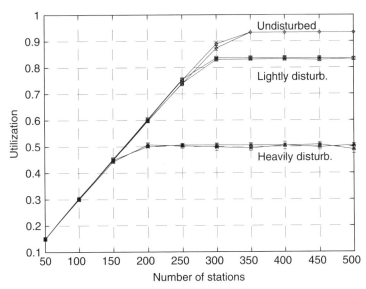

Figure 6.14 Average network utilization – basic ALOHA and basic polling protocols with rare transmission requests (average packet size: 1500 bytes)

offered network load of 313 to 314 network stations (784/2.5 = 313.6 – from Eq. 6.1).

$$C_N = n_{CH} \cdot \frac{S_p}{t_{TS}} \qquad (6.2)$$

C_N – total net capacity
n_{CH} – number of transmission channels
S_p – size of the segment payload (28 bytes)
t_{TS} – duration of a time slot (4 ms)

Network utilization in the lower load area also corresponds exactly to the total offered traffic. So, both protocols achieve an ideal utilization in the network without disturbances.

In the lightly disturbed network, there is about a 10% decrease in the available network capacity (Fig. 6.14). Accordingly, the saturation point moves left to 282/283 network station, which is also about 10% less than in the network without disturbances. In the heavily disturbed network, available network capacity and the saturation point decreases to 50% (saturation point at 156/157 network station). However, it can be concluded that in spite of data rate reduction in disturbed networks, network utilization maintains ideal behavior according to the available network capacity.

In the case of frequent transmission requests (simple data traffic model with average packet size of 300 bytes, Sec. 6.2.3), network utilization is lower for both ALOHA and polling reservation protocols (Fig. 6.15). In the network with the ALOHA access method, maximum utilization is achieved for 100 network stations (about 27%). Above 100 network stations, utilization decreases rapidly because of the increasing number of transmission demands caused by a higher number of arriving packets, which increases the number of collisions in the signaling channel.

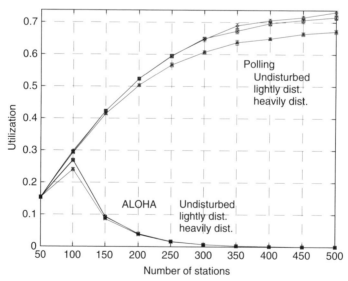

Figure 6.15 Average network utilization – basic ALOHA and basic polling protocols with frequent transmission requests (average packet size: 300 bytes)

As described above, the transmission channels are divided into time slots and a time slot can carry a transmission request (Fig. 6.11), which leads to a slotted ALOHA access method applied to the signaling channel. On the other hand, the maximum throughput of the slotted ALOHA protocol is 37% (Sec. 5.3.2), which means that a maximum of 37% of the transmission requests can be successfully sent to the base station. The duration of a time slot is 4 ms (Sec. 6.2.4), which means 250 time slots per second. So, if slotted ALOHA is applied to the signaling channel, a maximum of 92.5 requests can be successfully transmitted ($0.37/0.004 = 92.5$ according to Eq. 6.3).

$$r_S = G_{max} \cdot \frac{1}{t_{TS}} \tag{6.3}$$

r_S — number of successful requests (per second)
G_{max} — maximum throughput
t_{TS} — duration of a time slot (4 ms)

In the case of frequent transmission requests, the average packet size is 300 bytes (2400 bits), according to the simple traffic model. On average, 92.5 packets are transmitted per second, which amounts to a maximum of 222 kbps offered load in the network (Eq. 6.1), while the common net data rate is 840 kbps (15 channels with 56 kbps, including the signaling channel, Eq. 6.2). It results in a maximum of 26.43% network utilization, which confirms the simulation results (Fig. 6.15). Accordingly, in the case of rare requests (average packet size of 1500 bytes, 12,000 bits, according to the simple traffic model), the maximum offered load is 1110 kbps (Eq. 6.3), which is higher than the maximum net data rate in the network. Therefore, a nearly full network utilization – theoretical maximum – can be achieved in the case of rare transmission requests (Fig. 6.14).

The polling access method behaves much better than the ALOHA protocol in the network with frequent transmission requests (Fig. 6.15). However, a nearly full network utilization is not achieved. A larger number of network stations increase polling round-trip time and the stations have to wait longer to send the transmission requests. A request for only one packet can be transmitted each time, and this is the reason for the lower network utilization in the case of frequent requests and smaller packets.

If there are 400 stations in the network, polling round-trip time is 1.6 s (a request slot of 4 ms for each of 400 stations), according to Eq. 6.4.

$$t_{RTT} = n_{NS} \cdot t_{TS} \tag{6.4}$$

t_{RTT} — round-trip time of a polling message
n_{NS} — number of network stations
t_{TS} — duration of a time-slot (4 ms)

This means that a network station can send a packet (average size of 300 bytes, 2400 bit) within 1.6 s, which corresponds to its maximum offered traffic load of 1.5 kbps (Eq. 6.5).

$$l^{RTT}_{max} = \frac{P}{t_{RTT}} = \frac{P}{n_{NS} \cdot t_{TS}} \tag{6.5}$$

l^{RTT}_{max} — maximum network load per station under certain RTT
P — average packet size

In Eq. 6.1 $l = l^{RTT}{}_{max}$, the total network load amounts to 600 kbps for 400 stations, which is about 71% of the common net data rate (840 kbps). This network utilization is also evaluated by the simulation. On the other hand, in the case of rare requests, the maximum possible offered load per station in the network with 400 network stations is 7.5 kbps (every 1.6 s, a packet with average size 1500 bytes can be transmitted, Eq. 6.5). This is much higher than the average offered load of a network station (2.5 kbps), and therefore, the theoretically full network utilization can be achieved (Fig. 6.14).

The disturbances decrease the network utilization also in the case of frequent transmission requests (Fig. 6.15). However, the impact of disturbances is significantly lower than in the case of rare transmission requests. As mentioned above, in the case of a disturbed data segment, a whole user packet has to be retransmitted. Accordingly, the retransmission of smaller packets (300 bytes), occurring in the networks with frequent requests, occupies a smaller part of the network capacity than retransmission of the larger packets (1500 bytes). Therefore, networks with rare transmission requests are more affected by the disturbances than the networks with frequent requests.

6.3.1.3 Packet Delays

The following packet delays can be observed on the MAC layer:

- signaling delay,
- access delay, and
- transmission delay.

Signaling delay is defined as the time needed for the realization of the signaling procedure for a user packet. It is measured independently of the implemented access scheme and includes the time between packet arrival in the transmission queue of a network station (Fig. 6.7) and reception of acknowledgments from the base station (Fig. 6.16).

The access delay is measured from the time of the packet arrival until the start of the transmission. It includes the signaling delay and the waiting time, which is the time between reception of the acknowledgment from the base station and start of the transmission (Fig. 6.12). The transmission delay is the time between the packet arrival and the end of its transmission. It includes both signaling and waiting time, as well as the time needed for packet transfer through the network (Fig. 6.16).

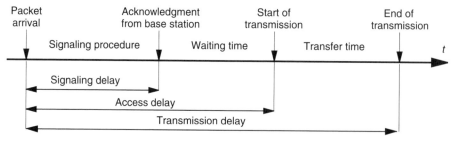

Figure 6.16 Packet delays

Signaling Delay

In the networks with rare transmission requests, the signaling delay is significantly shorter if ALOHA signaling protocol is applied than in the case of polling protocol (Fig. 6.17). On the other hand, the polling procedure causes a linear increase in the signaling delay according to the number of network stations (note, y-axis is presented in logarithmic scale).

If there are 50 stations in the network, a station receives a polling message from the base station every 50 time slots (or 200 ms, the duration of a time slot is 4 ms, Eq. 6.4) according to the round-robin procedure. If there are 500 stations, the t_{RTT} is 2000 ms. The packets arrive at the transmission queue of a network station randomly within the interval between two polling messages (RTT, Fig. 6.18). In the case of a network with rare requests, the average interarrival time (IAT) of the packets is 4.8 s (Tab. 6.1, Sec. 6.2.3).

If it is assumed that the packet arrivals are uniformly distributed within the RTT interval, the average signaling delay for polling protocol can be calculated in accordance with Eq. 6.6, where the t_{RTT} is given by Eq. 6.4. On average, a network station has to wait a half of the round-trip time for a polling message to transmit its request, which amounts to around 100 ms and 1000 ms in networks with 50 and 500 stations respectively. However, there is an additional time for receiving an acknowledgment from the base station (one time slot, Fig. 6.13), which additionally increases the signaling delay by 4 ms, as also

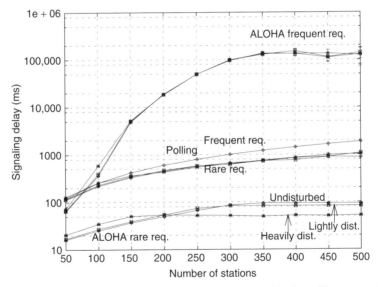

Figure 6.17 Mean signaling delay – basic ALOHA and basic polling protocols

Figure 6.18 Relation between round-trip time of polling messages and packet arrivals

confirmed by the simulation results (Fig. 6.17).

$$T_{\text{sig}} = \frac{t_{\text{RTT}}}{2} + t_{\text{Ack}} \quad (6.6)$$

T_{sig} – average signaling delay
t_{RTT} – round-trip time of a polling message
t_{Ack} – transmission time of an acknowledgment (4 ms)

In the case of ALOHA protocol, the signaling delay in the low load area is longer in disturbed networks than in the disturbance-free network. However, above the network saturation points (150–200, 250–300, 300–350 network stations in heavily, lightly and undisturbed networks respectively), the signaling delay in distributed networks is shorter. Above the saturation point, maximum network utilization is achieved and the transmission times of the packets increase, whereas the data throughput decreases (as is shown below, Fig. 6.22). Accordingly, the number of new transmission requests decreases because a new request can be sent after a packet is successfully transmitted. Therefore, the access delays in the high load area become shorter in disturbed networks than in the disturbance-free network.

In the networks with frequent transmission requests, polling protocol ensures significantly shorter signaling delays than ALOHA protocol (Fig. 6.17). Frequent transmission requests cause a higher number of collisions in the signaling channel and accordingly, a higher number of retransmissions, if ALOHA protocols are applied. Therefore, the signaling delays become extremely long.

In the case of polling, there is a nearly linear increase in the signaling delay. However, the signaling delay in the network with frequent requests also increases compared with the network with rare requests. There is the following reason for this behavior: transmission of smaller packets (300 bytes) is completed significantly faster compared to the large packets (1500 bytes), which makes possible the transmission of a request for the next packet, if any. Accordingly, the access and the transmission delays of the small packets (networks with frequent requests) consist mainly of the signaling delay, as is shown in Fig. 6.19. On the other hand, the IAT of the packets is significantly shorter in networks with frequent

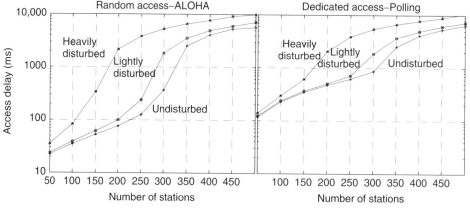

Figure 6.19 Mean access delay – basic protocols (average packet size: 1500 bytes)

requests (0.96 s) and there is a higher probability that a station has a new packet to transmit immediately after the previous packet is successfully transmitted. However, the station has to wait for the next polling message to transmit the new request.

If the new packet is ready immediately after the previous request (because the transmission is completed shortly afterwards), the network station has to wait longer for the new dedicated slot (almost the whole RTT) and the time between the packet arrival and the completion of the signaling procedure is increased. However, in the disturbance-free network, the maximum signaling delay cannot cross the round-trip time of the polling message, including the time needed for acknowledgment from the base station (Eq. 6.4); for example, the achieved signaling delay for 500 stations is 1875 ms and the polling round-trip time is 2000 ms, Eq. 6.6.

Access Delay

In the networks with rare requests and large packets (1500 bytes), the access delays are shorter if the ALOHA access method is applied (Fig. 6.19). The difference is more significant in the low network load area, below network saturation points. Above the saturation points (about 350, 250 and 200 network stations for disturbance-free, lightly and heavily disturbed networks respectively), ALOHA protocol still achieves shorter access delays, but the differences from the polling protocol are smaller.

In a low loaded network, a significant part of the access delay belongs to the signaling delay. Therefore, shorter signaling delays within ALOHA protocol for rare requests result in shorter access delays as well. However, above the saturation point at which the maximum network utilization is achieved, waiting time consumes a larger part of the access delay. The waiting time does not depend on the applied access method and increases proportionally with the network load. Therefore, the influence of the signaling delay and applied access methods decrease, and the access delays of ALOHA and polling protocols become closer. For the same reason, access delays in disturbed networks behave oppositely to the signaling delay and also remain longer in the high network load area.

On the other hand, the access delay in networks with frequent requests behaves in the same way as the signaling delay (Fig. 6.17), as shown in [Hras03], [HrasHa00], and [HrasHa01]. In both ALOHA and polling access protocols, the access delay depends mainly on the signaling delay, which is the reason for the same behavior.

Transmission Delay

Transmission delay includes the signaling and the waiting time, as well as the time needed for the packet transfer. The difference between the transmission and the access delays in highly loaded networks is very small for both random and dedicated access protocols (Fig. 6.20). Also, the shape of the curve for both transmission and access delays, which depend on the network load, remain the same.

A significant part of the transmission delay in the low network load area is caused by the signaling delay. On the other hand, the signaling takes a small part of the transmission delay in high loaded networks, particularly in the case of random access protocol. In the high loaded network, an almost full utilization is achieved and the waiting time for the beginning of a transmission increases significantly. This also raises the transmission delay, but does not have any influence on the signaling delay.

In the case of frequent requests (small packets 300 bytes), the difference between various packet delays is very small. The transfer time of small packets is relatively short compared

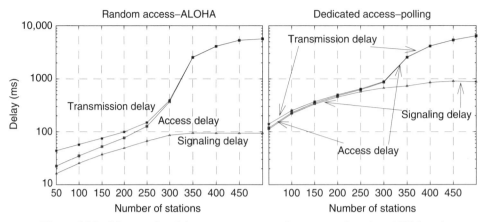

Figure 6.20 Mean packet delays – rare requests (average packet size: 1500 bytes)

with the time needed for the larger packets. On the other hand, networks with frequent requests do not achieve a nearly full utilization, which causes very short access delays, too. Therefore, the transmission delay depends mainly on the signaling delay, as shown in [Hras03], [HrasHa00] and [HrasHa01] as well.

The transmission delay behaves in the same way as the access delay in networks with disturbances. Of course, the transmission delays are longer than the access delays, but the curve shapes and their characteristic points remain the same (Fig. 6.21).

6.3.1.4 Data Throughput

The relative average data throughput is calculated as a ratio of the transmitted data and offered data rate of a network station. The data throughput follows the results achieved for network utilization. In networks with rare transmission requests, the maximum data throughput begins to decrease for 150, 250 and 300 network stations (Fig. 6.22), which

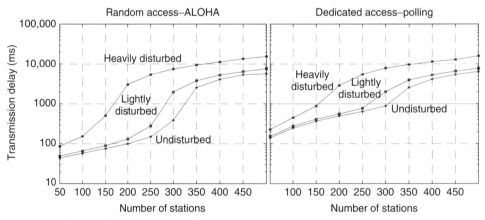

Figure 6.21 Mean transmission delay – rare requests (average packet size: 1500 bytes)

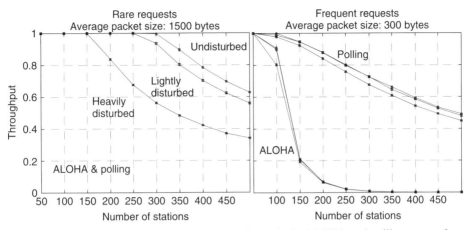

Figure 6.22 Average data throughput per station – basic ALOHA and polling protocols

are outlined as network saturation points in heavily and lightly disturbed networks and the undisturbed network respectively. The behavior of both random and dedicated access protocols remains the same as well.

The average data throughput in networks with frequent requests also behaves in the same way. Accordingly, the throughput decreases significantly above 100 network stations if ALOHA random access protocol is applied to the signaling channel. Dedicated polling protocol behaves better, but the decrease of the throughput is more significant than in the network with rare requests, which is also in accordance with the results for network utilization.

6.3.1.5 Conclusions

The purpose of the investigation of two basic reservation MAC protocols (random and dedicated access methods realized by slotted ALOHA and polling) is the validation of the simulation model and its elements, as well as the performance analysis of the basic protocol solutions. The calculations carried out in parallel (presented above) confirm the simulation results and prove the accuracy of the simulation model.

Two sets of parameters are used for traffic modeling, to represent networks with rare and frequent transmission requests (large and small user packets with average sizes of 1500 and 300 bytes respectively). It is shown that network performance depends strongly on this parameter set, which can be outlined as a suitable solution for the traffic modeling, ensuring protocol investigation and comparison under different traffic conditions. Noise scenarios applied within the disturbance model decrease the network performances by approximately 10% in lightly disturbed networks and by 50% in heavily disturbed networks, which provides a good basis for the observation of disturbance influence on the protocol and network performance as well.

A strong relationship between network utilization and data throughput is recognized for both protocol variants and all applied traffic and disturbance models. Access and transmission delays depend on the signaling delay in low network load area. However, in the highly loaded networks, they depend strongly on the entire network data rate. On the other hand, signaling delay indicates directly the efficiency of applied access protocol. The results evaluated for the signaling delay vary significantly in various network load areas under different disturbance conditions. It can be concluded that it is possible to evaluate the protocol performance by observing the network utilization and the signaling delay.

The signaling delays evaluated in the network using ALOHA-based reservation protocol are significantly shorter than with the polling access method, in the case in which transmission requests relatively seldom occur with accordingly fewer numbers of collisions in the signaling channel. However, if the collision probability increases (e.g. with increasing network load or number of subscribers in the PLC network), the advantage of the ALOHA-based protocol disappears. So, in the case of frequent transmission requests, the network applying ALOHA protocol collapses and polling has a significantly better performance.

6.3.2 Protocol Extensions

As shown above (Sec. 6.3.1), the basic reservation protocols behave differently under various traffic and network load conditions. In the case of random access protocol, network performance can be improved if the number of collisions appearing in the signaling channel is reduced. On the other hand, the disadvantages of the polling-based access method, applied to the signaling channel, can be improved by the insertion of a random component into the protocol, thereby decreasing the signaling delay in the low network load area [HrasHa01].

The basic reservation protocols can be extended in different ways, allowing for the combinations of various approaches (Sec. 6.1.4). In this investigation, we analyze the piggybacking access method, application of dynamic backoff mechanism, and extended random access principle.

6.3.2.1 Piggybacking

If the piggybacking access method is applied (e.g. [AkyiMc99], [AkyiLe99]), a network station transmitting the last segment of a packet can also use this segment to request a transmission for a new packet, if there is one in its packet queue (Fig. 6.7). The transmission request is not transferred over the signaling channel but is piggybacked within the last data segment. Accordingly, the application of piggybacking releases the signaling channel.

If a random access scheme is combined with piggybacking, the release of the signaling channel reduces the collision probability, thereby improving the delays and the throughput. In the case of the polling access method combined with piggybacking, the requesting station does not have to wait for a dedicated request, which would decrease the signaling delay and as a result, the data throughput is improved. A disadvantage of piggybacking is an overhead within data segments, which has to be provided for the realization of the piggybacking access method.

6.3.2.2 Dynamic Backoff Mechanism

In Sec. 5.3.2, we presented the principle of the dynamic backoff mechanism applied to the contention MAC protocol for stabilization of their performance. The dynamic backoff mechanism does not need a feedback information transmitted from the base station, which makes this mechanism suitable for application in PLC networks. To stabilize the performance of the signaling MAC protocols, it is also possible to apply a dynamic backoff mechanism.

There are several algorithms proposed for the dynamic change of the contention window (e.g. [DengCh00], [CameZu00], and [NatkPa00]). In this investigation, we apply a dynamic backoff mechanism without reset of the collision counter, introduced in [Hras03] and [HrasLe01], where an actual contention window is calculated from a basic (initial) contention window and a collision counter (Eq. 6.7).

$$CW[\text{TimeSlots}] = BCW \cdot CC \tag{6.7}$$

CW – Contention Window
BCW – Basic Contention Window
CC – Collision Counter

If a new packet arrives at a network station (its transmission queue), a transmission request is sent immediately, for example, in accordance with slotted ALOHA protocol. In the case in which the request was not successful, the collision counter (CC) is incremented by 1 (Fig. 6.23). The time for the request retransmission is then randomly computed from the

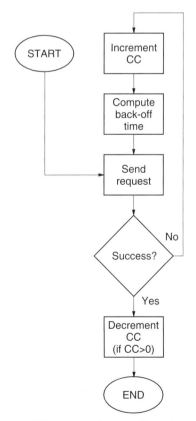

Figure 6.23 Dynamic backoff mechanism

actual contention window (CW). So, the contention window is increased every time a collision occurs.

After a successful request transmission, the collision counter is not immediately set to zero, such as in various backoff mechanisms, but it is decremented by 1, if CC is greater than zero.

6.3.2.3 Extended Random Access

A further possibility for performance improvement of basic reservation MAC protocols is the application of so-called extended random access, proposed in [HrasLe02], which is realized over data channels to ensure an additional possibility for transmission of the requests. However, the data channels can only be used for signaling if they are currently free (not used for any data transmission). Thus, the collision-free data transmission, provided by the reservation MAC protocols, is kept.

Network stations are not able to get any information about the channel occupancy in the uplink transmission direction, for example, if another station simultaneously starts to send its data. Therefore, it is necessary to provide any information about the channel occupancy in the uplink, ensuring collision-free data transmission. On the other hand, if the base station has the information that the data channels are busy or free during certain time slots, then accordingly it is able to provide this information to the network stations (e.g. by broadcasting it in the downlink informing the stations about the next uplink time slot). So, the extended random access to the collision-free data channels can be realized by broadcasting an additional channel occupancy information, similar to the ISMA protocol (Sec. 5.3.2).

6.3.2.4 Analysis of Extended Protocols

To investigate the impact of the protocol extensions on the network performance, we implemented piggybacking, dynamic backoff mechanism, and extended random access (all extensions are described above) within both basic protocols ALOHA and polling (Sec. 6.3.1), creating two so-called extended protocol solutions – extended ALOHA and extended hybrid polling–reservation protocols. With the implementation of the extended random access within the polling protocol, we could make a hybrid polling–protocol solution. So, the hybrid-polling solution includes an additional random component. Organization of the signaling channel remains the same as in the case of basic ALOHA and polling reservation protocols.

Network Utilization

In networks with rare transmission requests (large packets), there is no difference between network utilization achieved for basic solutions (Fig. 6.14) and for extended ALOHA and extended hybrid-polling protocols. However, a big improvement can be recognized with extended ALOHA protocol in the case of frequent transmission requests compared with basic ALOHA (Fig. 6.24). As expected, piggybacking improves network utilization in the high network load area, and dynamic backoff mechanism stabilizes network utilization. On the other hand, the usage of data channels for signaling increases network utilization in the low load area. However, the maximum network utilization is about 73%, achieved in the disturbance-free network, which is lower than the theoretical maximum.

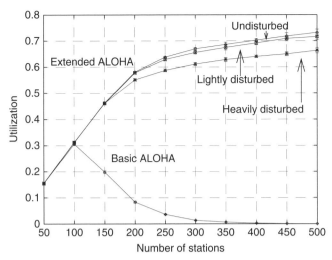

Figure 6.24 Average network utilization – extended ALOHA protocol – frequent requests (average packet size: 300 bytes)

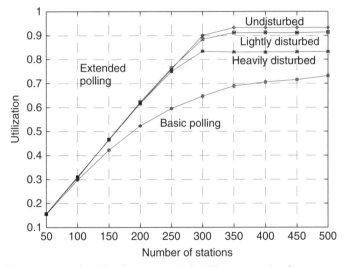

Figure 6.25 Average network utilization – extended polling protocol – frequent requests (average packet size: 300 bytes)

Extension of basic polling protocols improves the performance significantly, achieving a nearly full network utilization, as also in the case of frequent transmission requests (Fig. 6.25). This improvement is reached thanks to the piggybacking access method, which takes the most request transmissions in the high network load area. In this case, the network stations do not have to wait for polling messages to send the requests because they can use piggybacking, avoiding the negative influence of long round-trip times for the polling messages.

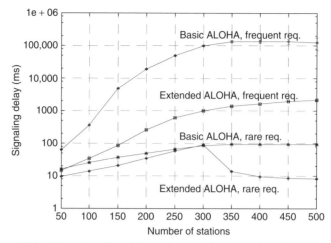

Figure 6.26 Mean signaling delay – basic and extended ALOHA protocols

Signaling Delay
Extended ALOHA protocol ensures shorter signaling delays than basic ALOHA protocol in both cases – rare and frequent transmission requests (Fig. 6.26). Below 300 stations in the network (near the network saturation point), the shorter delays are caused by usage of free data channels for the signaling. Above the saturation point, the effect of piggybacking can be observed, which significantly decreases the signaling delay. This improvement is even more significant in networks with frequent requests also due to both piggybacking and signaling over data channels. The stabilization of the signaling delay is achieved by application of the dynamic backoff mechanism.

Usage of data channels for signaling and dynamic backoff mechanisms applied to the extended polling protocol decreases the signaling delay below network saturation point as well (Fig. 6.27). Above the saturation point, the effect of piggybacking is again

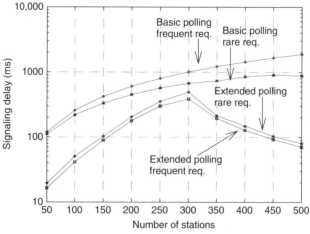

Figure 6.27 Mean signaling delay – basic polling and extended hybrid polling

recognized. Signaling delay for networks with frequent requests using basic polling protocol, are longer than for rare requests (see Sec. 6.3.1). Otherwise, with the extended polling protocol, the small packets have slightly shorter signaling delays, because of more possibilities for the request transmission (piggybacking, usage of data channels, and signaling channel).

Signaling Delay in Disturbed Networks
A comparison of extended ALOHA and extended polling protocols operating in disturbed networks is presented in Figs. 6.28 and 6.29 respectively. In the case of rare requests, extended ALOHA behaves better than extended polling and achieves a maximum signaling delay of 90 ms. Extended polling achieves a maximum signaling delay of 500 ms. On the

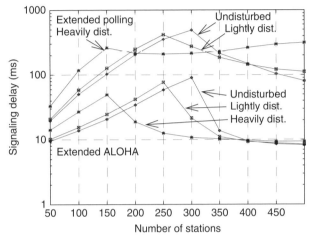

Figure 6.28 Mean signaling delay – extended ALOHA and hybrid-polling protocols – rare requests (average packet size: 1500 bytes)

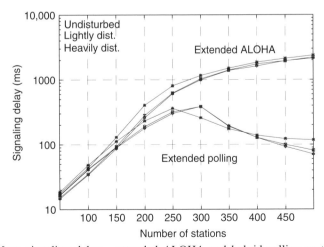

Figure 6.29 Mean signaling delay – extended ALOHA and hybrid-polling protocols – frequent requests (average packet size: 300 bytes)

other hand, signaling delay in networks with frequent requests remains under 400 ms if extended polling is applied. Extended ALOHA achieves significantly longer signaling delays (over 1 s). So, the extended protocols keep the features of their basic protocol solutions in both networks with rare and frequent transmission requests (Sec. 6.3.1)

It can also be concluded that extended protocols behave the same as their basic protocol solutions in networks operating under disturbances. Signaling delay in low network load area is longer in more disturbed networks, but near the network saturation point and beyond the saturation point it becomes shorter than in the networks without disturbances. This behavior is already explained in Sec. 6.3.1.

6.3.2.5 Conclusions

The usage of data channels for signaling improves network performance significantly in the low network load area, as well as the piggybacking access method in the high loaded networks. By the application of the dynamic backoff mechanism, ALOHA protocol can be stabilized in the case of frequent transmission requests. Extended ALOHA protocol keeps the features of its basic variant and provides shorter signaling delays than extended polling in the network with rare transmission requests.

Extended polling behaves better in the case of frequent requests and with it keeps the features of its basic variant as well. Different from extended ALOHA, extended polling protocol always provides a theoretical maximum network utilization and acceptable signaling delay in the low and high load areas. However, the delays are still too long in the middle load area (over 100 ms, which is not suitable for time-critical services), near network saturation point, particularly in networks with rare transmission requests.

6.3.3 Advanced Polling-based Reservation Protocols

The comparison of the extended reservation protocols shows some advantages of polling-based access methods (Sec. 6.3.2). However, the signaling delay achieved by extended polling is still too long for the realization of time-critical services. The signaling delay can be additionally reduced by decreasing the round-trip time of polling messages. This can be ensured by application of the active polling method (Sec. 5.3.3), as shown in [HrasLe02a] and [HaidHr02]. A further reduction of the round-trip time can be achieved by dynamic association of the network station [KellWa99] in a list of so-called active stations. Finally, the polling procedure realizes continuous communication between the base station and network stations, which is useful for various network control tasks, thereby improving the overall network performance (e.g. application of different scheduling strategies, fault management, etc.). A disadvantage of polling-based protocols is a relatively high realization complexity compared with ALOHA protocols. However, because of the current developments regarding microcontrollers and signal processors, implementation of polling-based protocols does not seem to be difficult.

6.3.3.1 Signaling Protocol Based on Active Polling

With the increasing number of stations in the network, the round-trip time of polling messages (Sec. 6.3.1) increases. A station sends a request only for the transmission of a packet and after an acknowledgment from the base station, it waits for the transmission. Just after a successful packet transmission, the station can transmit a request for the next packet, if there are any in its queue. In between, the dedicated request slots for the station remain unused. Active polling (Sec. 5.3.3) is used to avoid this situation and to reduce the delays in the network. The idea of active polling is that only so-called active network stations are polled while other stations are temporarily excluded from the polling circle. The active network stations are potential data transmitters and the other stations do not currently send any data.

Active polling can also be implemented within the signaling MAC protocol in the considered polling-based reservation protocols [HrasLe02a]. However, in this case the active stations are the network stations, which are potential transmitters of the transmission requests. The polling message determining the dedicated request slot is not transmitted to sending and waiting network stations until they complete the transmission, because they do not send a new request until the end of their transmissions. The polling messages are only sent to the stations that currently do not transmit data or that are not waiting for a transmission right because they are potential transmitters of new transmission requests. In this way, the signaling delay can be additionally reduced.

Active polling can be implemented within the basic polling protocol, as well as within the extended polling. On the other hand, both basic and active polling can be combined with piggybacking as well. Various polling protocol solutions, investigated below, are represented in Tab. 6.3.

6.3.3.2 Performance Evaluation of Polling Based Reservation MAC Protocols

Figure 6.30 presents the network utilization in networks with frequent transmission requests for the investigated variants of polling-based reservation MAC protocols. Application of active polling increases the network utilization, but only very slightly. On the other hand, application of piggybacking increases network utilization significantly and achieves the theoretical maximum value (Sec. 6.3.1). Accordingly, a combined usage of active polling and piggybacking access methods ensures a nearly full network utilization as well. In the case of rare transmission requests, all investigated variants of polling-based reservation protocols achieve the maximum possible network utilization.

Table 6.3 Investigated polling-based reservation MAC protocols

Protocol	Description
(Basic) polling	Dedicated access to the signaling channel realized by polling
Active polling	Only active stations are polled (potential transmitters of requests)
Polling with piggybacking	Basic polling with piggybacking
Active polling with piggybacking	Active polling with piggybacking
Extended polling	Basic polling with piggybacking, extended random access, and dynamic backoff mechanism
Extended active polling	Active polling with piggybacking, extended random access, and dynamic backoff mechanism

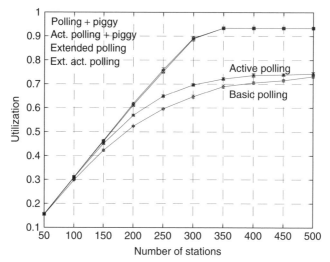

Figure 6.30 Average network utilization – polling-based reservation MAC protocols – frequent requests (average packet size: 300 bytes)

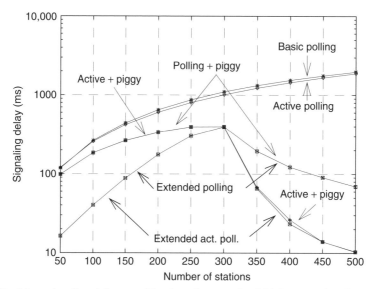

Figure 6.31 Mean signaling delay – polling-based reservation MAC protocols – frequent requests (average packet size: 300 bytes)

Figure 6.31 presents simulation results for mean signaling delay in the network with frequent transmission requests using different polling-based reservation protocols. The improvement achieved by the usage of active polling compared with basic polling protocol is hardly noticeable. On the other hand, significantly shorter signaling delays can be observed in the high network load area, for both active and basic polling combined with piggybacking. However, the combination of active polling and piggybacking, as well as both extended basic polling and extended active polling achieve the best results in highly loaded network.

In low and middle load areas, both combinations of basic and active polling with piggybacking ensure shorter signaling delays than protocols without piggybacking. The best results in the low load area are achieved by both extended protocols, which additionally apply the extended random access to data channels to be used for signaling including the dynamic backoff mechanism and piggybacking. However, near the network saturation point, (300 stations) the extended protocols do not improve the performance compared with basic and active polling with piggybacking.

In networks with rare transmission requests, the signaling delay decreases in the high network load area if active polling or piggybacking are applied (Fig. 6.32). The combination of active polling and piggybacking achieves again the best results, as is the case in the network with frequent requests. However, the influence of active polling is more significant if longer user packets are transmitted. The reason for this is a correspondingly longer absence of dedicated request slots for the stations sending or waiting for access rights, which is caused by the longer transmission times necessary for larger packets. This decreases the general round-trip times of the polling messages and reduces the signaling delay as well. Reduction of the signaling delay in the low network load area is also achieved by application of both extended protocols.

Generally, it can be concluded that the combination of active polling and piggybacking improves the network utilization, ensuring the theoretical maximum value in both networks with frequent and rare transmission requests. This protocol combination reduces the signaling delay significantly in high network load area. However, the signaling delay in the middle load area is still longer than in the case of protocols with random access in networks with rare requests (e.g. extended ALOHA, Sec. 6.3.2).

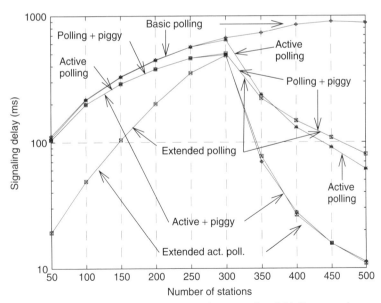

Figure 6.32 Mean signaling delay – polling-based reservation MAC protocols – rare requests (average packet size: 1500 bytes)

6.3.3.3 Two-step Reservation Protocol

From the investigation of polling-based reservation MAC protocols presented above, we can conclude that the number of active network stations in the medium network load area is still high, giving rise to longer round-trip times of polling messages. Accordingly, the reduction of the signaling delay is not so significant, as is the case in the high network load area where the network stations are mainly in the waiting state and belong to the group of inactive stations. Therefore, a reduction of signaling delays in medium network load area is only possible if the number of active stations is decreased. This can be ensured by a division of the polling procedure into two phases, as is proposed in [Hras03] and [HrasLe02b]:

- *prepolling phase* – used for estimating the active network stations, and
- *polling phase* – including standard polling procedure of the active stations.

For the realization of the two-step reservation procedure, downlink signaling slots are divided into three fields (Fig. 6.33). The first field is reserved for transmission of a so-called prepolling message, which specifies a group of network stations that can set a prerequest in the next uplink signaling slot. The other two fields in the downlink are used according to the standard polling procedure (Sec. 6.3.1); for polling messages addressing a network station to send a transmission request in the next uplink signaling slot, and for acknowledgments from the base station containing information about the access rights.

In the uplink there are a number of so-called prerequest microslots and a request field. Each of the microslots is reserved for a network station that is a member of a group specified in the prepolling message from the previous time slot in the downlink. The microslots are created to take up a minimum part of the network capacity, allowing just the transfer of transmission indications for a number of network stations, simultaneously and without collisions. The request field is used for the request transmission after a station is polled in the previous downlink slot (standard polling procedure).

The order of events within the two-step reservation procedure is presented in Fig. 6.34. After a station receives a prerequest polling message that addresses its group, it uses one of the prerequest microslots in the next uplink signaling slot to set a prerequest. Note that within a group of stations there are dedicated prerequest microslots for each of them and this ensures a contention-free transmission of the prerequests. After that, the base station transmits a polling message to the requesting station.

The base station can receive multiple prerequests, but it can send only one polling message within a signaling slot. Therefore, there is a need for the scheduling of arrived prerequests. Accordingly, that can delay the transmission of polling messages to the stations

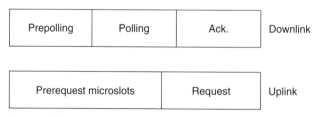

Figure 6.33 Slot structure for two-step reservation procedure

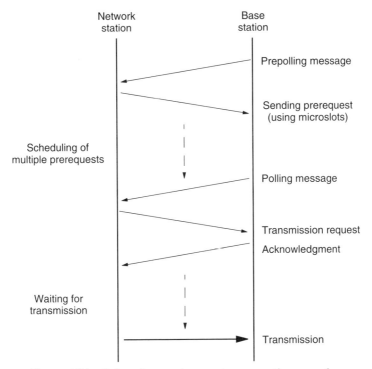

Figure 6.34 Order of events in two-step reservation procedure

that already sent the prerequests. After a station receives a polling message, it transmits a request in the next uplink time slot, such as in one-step reservation procedures. After that, an acknowledgment from the base station follows, which defines the access rights.

In the two-step reservation protocol, there exists the possibility that no other station will send a prerequest. In this case, and if the base station has already scheduled all previous received prerequests, no station is polled, and the request field in the next uplink signaling slot remains unused (Fig. 6.35). Accordingly, the two-step protocol can be extended to allow random access to the empty request slots, making a hybrid-two-step protocol. In this case, the whole reservation procedure is avoided, thereby decreasing the signaling delay. Collisions between randomly realized multiple requests are possible, but only if the request fields are free for random access and if there was no polling message in the previous time slot. After the collisions occur, access to the medium is carried out in accordance with the basic two-step reservation method.

Generally, two-step or hybrid-two-step reservation protocols can also be extended by implementation of the following features:

- piggybacking,
- extended random access,
- dynamic backoff mechanism – for the hybrid part of the two-step procedure (access to the free request slots and free data channels), and
- active polling, applied to the first protocol phase.

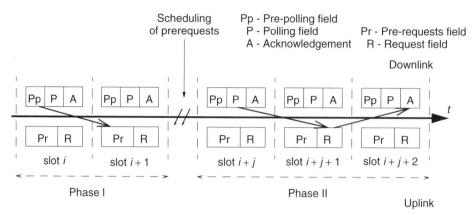

Figure 6.35 Two-step reservation protocol

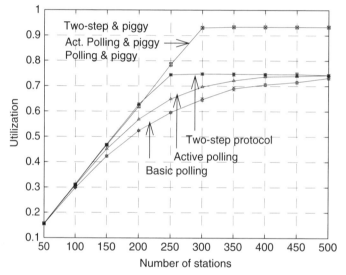

Figure 6.36 Average network utilization – two-step reservation protocol – frequent requests (average packet size: 300 bytes)

6.3.3.4 Performance Evaluation of Two-step Reservation MAC Protocol

A two-step reservation protocol achieves the maximum possible network utilization in the networks with rare transmission requests, as is also concluded for other investigated protocol variants. In the case of frequent requests, the two-step protocol achieves a slightly higher utilization than active polling and basic polling protocols (Fig. 6.36). The active polling method reduces the round-trip time of polling messages by addressing only potential requesting stations. However, it is possible that nonrequesting stations, without data to send, are also polled. The two-step procedure is more effective, especially in the middle

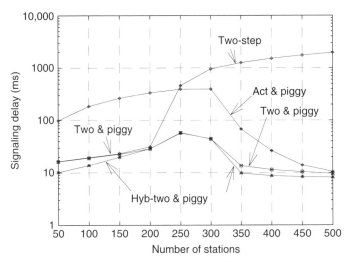

Figure 6.37 Mean signaling delay – two-step protocol – frequent requests (average packet size: 300 bytes)

load area, because it only addresses the requesting stations, which are determined during the first protocol phase. The theoretical maximum network utilization can be achieved using a combination of the two-step protocol and piggybacking.

In the network with frequent transmission requests, the two-step reservation procedure decreases the signaling delay significantly in the low network load area (Fig. 6.37). However, in the middle and high load areas, active polling with piggybacking (the best variant of polling-based protocols) behaves much better. This disadvantage of the two-step protocol can be solved by the usage of piggybacking (Two & Piggy), which reduces the signaling delay in the entire investigated load area.

The insertion of the hybrid component into two-step reservation protocols additionally decreases the signaling delay in low and high load areas. In this case, the request slots are mainly randomly accessed and the two-step reservation procedure is avoided, which shortens signaling delay. In the low load area, it is caused by a relatively small number of network stations and in the high load area, it is caused by piggybacking, which takes most of the requests and releases the signaling channel.

In the case of rare transmission requests (Fig. 6.38), the two-step reservation procedure provides a shorter signaling delay than active polling with piggybacking below 400 stations in the network. On the other hand, the two-step procedure, extended by piggybacking, behaves better in the entire investigated load area.

The hybrid-two-step protocol with piggybacking decreases the signaling delay significantly and keeps it relatively constant. The achieved results for the hybrid-two-step protocol are better than for any ALOHA-based access method investigated under similar conditions. So, it can be concluded that the hybrid-two-step reservation protocol with piggybacking behaves better than any investigated one-step protocol in networks with both rare and frequent transmission requests.

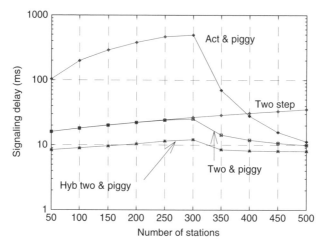

Figure 6.38 Mean signaling delay – two-step protocol – rare requests (average packet size: 1500 bytes)

6.4 Error Handling in Reservation MAC Protocols

In the investigation of signaling MAC protocols presented above (Sec. 6.3), we analyze two basic protocol solutions, ALOHA and polling, several possibilities for improvement of their performance by application of various protocol extensions, and advanced polling-based reservation protocols. It can be concluded that the polling protocols, particularly the two-step protocol, achieves the best performance.

In this section, we analyze the implementation of various error-handling mechanisms within reservation MAC protocols. Because of a more complex signaling procedure in polling-based protocols (active polling, two-step), we first investigate the possibilities of protecting the signaling procedure (Sec. 6.4.1). Application of ARQ mechanism, used for protection of the data flow in a network, within reservation MAC protocol is discussed in Sec. 6.4.2. Finally, in this section, we investigate the possibilities of integrating ARQ mechanisms within per-packet reservation MAC protocols (Sec. 6.4.3).

6.4.1 Protection of the Signaling Procedure

6.4.1.1 Protecting Active Polling Signaling Procedure

The application of active polling signaling procedure (Sec. 6.3.3) calls for additional error protection of the signaling procedure. If the base station receives a request from a network station, it does not send a new polling message to the requesting station before the transmission is finished. With it, the station is temporarily excluded from the polling circle until it completes the transmission. However, if the acknowledgment message from the base station has been disturbed, the requesting station does not receive information about its access rights and it does not start the transmission. This can cause the effect that this network station would never again be polled.

To avoid this situation, the following mechanism has to be implemented within the active polling protocol [HrasLe02c]: if a station does not start sending data at a specified moment (already reserved for its transmission), the base station recognizes it and transmits an extra polling message to the affected network station, which allows repetition of the transmission request. Thus, if a requesting network station does not receive an acknowledgment because it is disturbed, then it is ensured that it will not be excluded from the polling cycle for a longer time period.

6.4.1.2 Protected Two-step Protocol

Compared with one-step procedures, the usage of the two-step reservation procedure increases error probability, because there are more signaling messages transmitted between network stations and the base station. If a signaling message is missed, the whole reservation procedure has to be repeated, which decreases the network performance. Therefore, the network stations continuously set the transmission prerequests into permitted prerequest slots until the reservation procedure is finished. So, if one of the signaling messages is disturbed, the reservation procedure does not have to be repeated from the beginning. However, a multiple prerequest for a data packet could cause unnecessary reservation of the transmission capacity. To avoid this situation, a mechanism has to be implemented, to recognize and to avoid multiple reservations. Both repetition of the prerequests and avoidance of multiple reservations belong to the mechanism that protects the two-step signaling procedure [Hras03], [HrasLe02c].

Average network utilization in both networks applying the protected two-step reservation protocol is the same as in other efficient MAC protocols, achieving the maximum possible value; for example, such as ALOHA in the case of rare transmission requests (Fig. 6.14) and extended polling in the case of frequent requests (Fig. 6.25). Also in the disturbed networks, network utilization remains the same as with one-step protocols. Signaling delays in disturbed networks, applying the protected two-step protocol, behave in the same way as in the networks with one-step protocols as well [HrasLe02c]. Thus, we can conclude that the protected two-step protocol is not worse compared with the one-step protocols in the context of their usage in networks operating under unfavorable noise conditions, such as PLC.

6.4.1.3 Fast Re-signaling Procedure

In the previous investigations of different protocol variants, only a simple mechanism for packet retransmission was implemented in the case of erroneous transmissions. Network stations receive an acknowledgment from the base station if a packet is successfully transmitted. Otherwise, the affected station has to repeat the whole signaling procedure to retransmit the packet. To avoid the signaling repetition and to improve the protocol performance in disturbed networks, a fast re-signaling procedure is implemented to ensure automatic allocation of the network capacity for the necessary packet retransmission. So, in the case of an erroneous packet, the base station sends a negative acknowledgment to the affected network station, including an allocation message that contains the information about the access right for the packet retransmission.

Application of the fast re-signaling procedure reduces the signaling delay, and is particularly visible in heavily disturbed networks (Fig. 6.39). On the other hand, in an undisturbed network, fast re-signaling never runs and the signaling delay does not change.

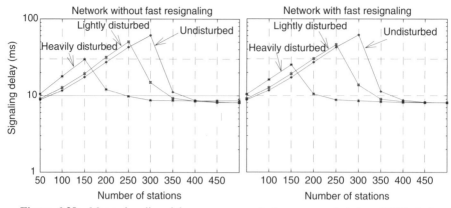

Figure 6.39 Mean signaling delay – rare requests (average packet size: 1500 bytes)

6.4.2 Integration of ARQ in Reservation MAC Protocols

As described above, in the case of reservation MAC protocols, a network station starts transmission of data segments belonging to a user packet (e.g. IP packet) by using a particularly allocated portion of the transmission resources. After a network station starts transmitting the data segments, it can happen that one or more segments are disturbed. In previous investigations, simple retransmission of the whole packet is applied if at least one segment of the packet is disturbed. However, in communications systems with higher BER, it is more efficient to retransmit smaller data units (Sec. 5.2.1). Therefore, ARQ is applied to retransmit erroneous segments and not the whole packet.

In the case of Go-back-N ARQ mechanism, the base station has knowledge of the number of requested segments and can discover if there are some erroneous or missing data segments on the receiving side. In this case, it sends a negative acknowledgment (NAK) to the sending station, including the sequence number of the last received segment. Thus, the sending station has to retransmit only the data segments with the higher sequence number. If the Selective-Reject ARQ mechanism that achieves the best performance from among different ARQ mechanisms is applied, the sending station retransmits only the erroneous data segment. Each of the ARQ variants, described in Sec. 4.3.4, can be applied together with reservation MAC protocols.

However, because of the applied per-packet reservation method, the affected station is not able to retransmit all disturbed data segments within the previously reserved transmission turn. It happens because a station receives the right only to send for the requested number of data segments and it is possible that another station will start to send immediately afterwards (Fig. 6.40). Therefore, the network station has to repeat the

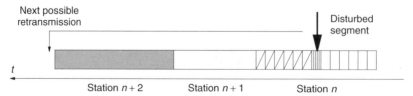

Figure 6.40 ARQ and per-packet reservation principle

transmission request for the disturbed packet. To avoid the repetition of the whole signaling procedure, NAK can be specified to also include the information about the access rights, such as in Fast Re-signaling procedure, as described above.

To reduce the number of ARQ related signaling messages to a minimum and also to decrease the network load caused by the ARQ signaling, the following procedure can be adopted: an ACK (positive acknowledgment) is sent only after a whole user packet is successfully received. In between, the NAK messages are sent to the sender only in the case of corrupted or missing data segments.

6.4.3 ARQ for Per-packet Reservation Protocols

6.4.3.1 ARQ-plus Mechanism

In the case of the ARQ mechanism described above, a network station that has to retransmit a number of data segments (all succeeding data segments after a disturbed segment, Fig. 6.40) interrupts the transmission and the rest of the already allocated network capacity remains unused. These transmission gaps can be avoided by application of a so-called ARQ-plus mechanism, as shown in Fig. 6.41 [HrasLe02c]. In the case of an erroneous data segment, all succeeding segments have to be retransmitted as in the case of the ARQ mechanism described above, but the retransmission can start immediately. With it, the transmission gaps are kept as small as possible.

To ensure immediate retransmission, additional data slots have to be allocated to the affected network station (shift). The same number of data slots has also to be calculated for other network stations that are possibly waiting for the transmission, ensuring a correct collision-free data transmission. The reallocation information containing an exact shift value has to be included in the NAK message. Sometimes, the allocated transmission time for a station has to run out before it can receive a NAK from the base station (the next station has already started to send). In this case, application of the ARQ-plus mechanism is not possible and the retransmission proceeds according to the simple ARQ mechanism (Fig. 6.40).

6.4.3.2 ARQ-plus without Shifting

The ARQ-plus mechanism improves the network utilization and shortens the transmission delays. However, the reallocation of already reserved transmissions (shifting) can cause problems in a network operating under hard disturbance conditions, such as PLC. A reallocation message sent by the base station can also be disturbed, even selectively.

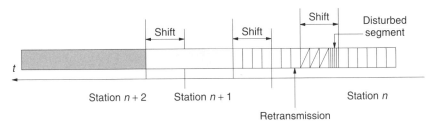

Figure 6.41 ARQ-plus mechanism

This means that it can happen that some stations already waiting for a transmission receive the reallocation message and other stations do not receive the message. It causes de-synchronization of the access to the medium, which leads to unwanted collisions decreasing the network utilization.

To avoid this situation, the ARQ-plus mechanism should be implemented without shifting. In this case, a station retransmitting data segments uses the reserved capacity for a number of segments to be retransmitted (Fig. 6.41). However, the reserved network capacity cannot be used for all data segments (because of the retransmissions, the number of segments to be transmitted is higher than originally reserved) and an additional reservation for the remaining segments is carried out according to the simple ARQ mechanism. The additional reservation is carried out according to the fast re-signaling procedure. In this way, network utilization remains such as in the ARQ-plus mechanism and the transmission time of affected packets becomes longer, but is still shorter than with the simple ARQ mechanism, as shown below.

6.4.3.3 Simulation Results

Figure 6.42 presents the average network utilization in networks with both rare and frequent transmission requests, using the simple packet retransmission for a two-step protocol. In Fig. 6.43, the results for networks applying Go-back-N ARQ mechanism are presented for comparison.

It can be concluded that application of the ARQ mechanism improves network utilization significantly. The improvement is especially visible if the networks with larger user packets are considered; 83 to 89% in lightly disturbed networks and 50 to 73% in heavily disturbed networks. In the case of smaller packets, the improvement is approximately 91 to 92% in lightly disturbed networks and 83 to 88% in heavily disturbed networks.

Network utilization is further increased by the application of ARQ-plus mechanisms (Fig. 6.44); ARQ-plus with shifting and ARQ-plus without shifting. In the case of larger user packets, the utilization of 92% is achieved in lightly disturbed networks and 81% in heavily disturbed networks. For the smaller user packets, the network utilization saturates

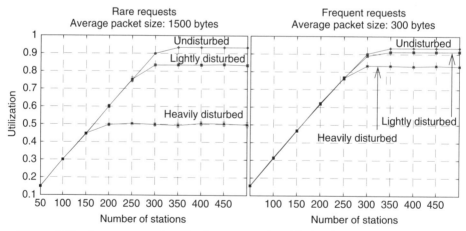

Figure 6.42 Average network utilization – networks with simple packet retransmission

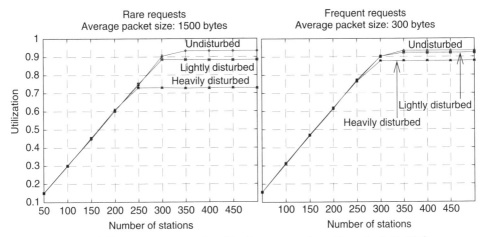

Figure 6.43 Average network utilization – networks with go-back-N ARQ

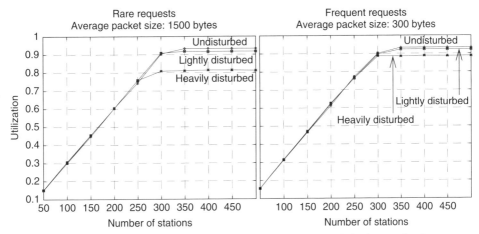

Figure 6.44 Average network utilization – networks with ARQ-plus mechanisms

to the maximum possible (about 93%) in lightly disturbed networks and to 90% in heavily disturbed networks.

The application of ARQ and ARQ-plus mechanisms improves the transmission delay significantly, as shown in Fig. 6.45. As expected, the network using ARQ-plus mechanism, which exploits possible retransmission gaps, achieves the shortest transmission delays. The ARQ-plus mechanism without shifting (ARQ + WS), achieves shorter transmission delays than simple ARQ mechanism in low loaded networks. However, the transmission delay remains longer than in the case of the ARQ-plus mechanism with shifting.

With the increasing network load, the transmission delay achieved in the network with the ARQ-plus mechanism without shifting comes close to the delay achieved by a simple ARQ. Beyond 200 stations in the network, the delays have practically the same value. Thus, application of the ARQ-plus mechanism without shifting ensures good network utilization (the same as ARQ-plus with shifting), but the transmission delay remains

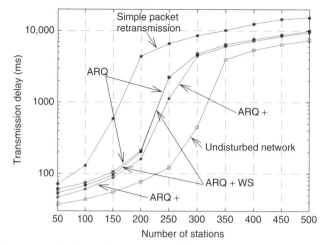

Figure 6.45 Mean transmission delay of user packets – networks with rare requests (average packet size: 1500 bytes)

almost the same as with the simple ARQ. However, the difference between transmission delays achieved by the simple ARQ and ARQ-plus mechanisms is small.

If the networks with small packets (frequent requests) are considered, the behavior of the transmission delay remains the same as is presented in Fig. 6.45. However, the transmission delays of larger packets are generally longer and the impact of the applied error-handling mechanisms is much higher as well [HrasLe02c].

6.5 Protocol Comparison

In previous sections, we investigated several protocol solutions for the signaling MAC protocols and for various protocol extensions. It is concluded that the two-step protocol achieves better performance than the so-called one-step protocols – ALOHA and polling-based solutions. The aim of the investigation in this section is a direct performance comparison of two-step and one-step reservation MAC protocols. For this purpose, extended ALOHA, extended active polling and extended hybrid-two-step protocols are investigated. To ensure a fair protocol comparison, we analyze the required slot structure in the signaling channel for realization of each investigated protocol (Sec. 6.5.1). This investigation is carried out with application of multimodal traffic models (Sec. 6.2.3), used for specification of a traffic mixture representing nearly realistic behavior of different network users (Sec. 6.5.2). Finally, the achieved simulation results (Sec. 6.5.3) are discussed in Sec. 6.5.4 in the context of realization of QoS for various telecommunication services in two-step protocol.

6.5.1 Specification of Required Slot Structure

6.5.1.1 Extended Hybrid-Two-step Protocol

In the specification of the network and simulation models (Sec. 6.2.4), we assume that a time slot of the implemented OFDMA/TDMA scheme has a duration of 4 ms and carries

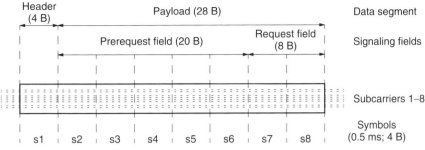

Figure 6.46 Realization of prerequest microslots

a data segment with a size of 32 bytes. Four bytes are reserved for the segment header and the remaining 28 bytes belong to the segment payload. The time-slot structure is the same for both signaling and data channels. If it is assumed that each transmission channel contains 8 subcarriers, the prerequest microslots needed for the two-step protocol can be realized within the uplink part of the signaling channel, as presented in Fig. 6.46. The header occupies 4 bytes, a request field 8 bytes, and the remaining 20 bytes can be used for the realization of the prerequest microslots, needed for the two-step reservation procedure.

If we assume that the duration of an OFDM symbol, including the payload and the guard symbol extension, can be set to 0.5 ms (Sec. 4.2.1), a data segment consists of 8 symbols, each carrying 4 bytes of information. Thus, 1 symbol is reserved for the segment header, 2 symbols are needed for the request field and 5 symbols within a signaling time slot can be used for the realization of the prerequest microslots (s2–s6). If each symbol is used as a microslot, there can be 5 prerequest-slots. If each subcarrier is used for 1 microslot, it is possible to create 40 microslots within the signaling time slot (5 symbols each with 8 subcarriers). The microslots are realized to occupy the minimum possible network resources and they just ensure a collision-free transmission of indications (prerequests) that a station has some data to send.

6.5.1.2 Extended ALOHA and Extended Active Polling

For realization of ALOHA reservation procedure, there is a request field in the uplink part of the signaling channel in every time slot, which can be used for the request transmission (Fig. 6.47). After a successful request (e.g. there was no collision with requests from other network stations), the base station transmits an acknowledgment in the downlink direction in the next time slot. In accordance with the slot structure, presented in Fig. 6.46, it can be concluded that it is possible to realize more than one request field within a time slot. Therefore, to ensure a fair comparison between investigated protocols, we assume that four transmission requests can be realized within a time slot, which is ensured by so-called request minislots (Fig. 6.47). The number of acknowledgments per time slot is set to four, as well.

In the case of polling, network stations can transmit their requests after they were polled in the previous time slot (Fig. 6.48). For this investigation, it is also assumed that the request field is divided into four minislots, such as in the case of ALOHA protocol, and that the base station can poll four network stations within a time slot.

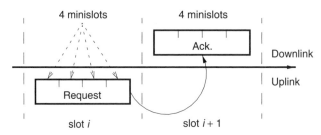

Figure 6.47 Slot structure for ALOHA protocol

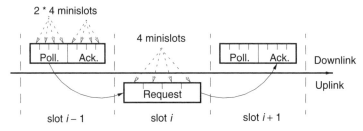

Figure 6.48 Slot structure for polling protocol

6.5.2 Specification of Traffic Mix

To specify a traffic mix to be used for the protocol comparison, we assume that 70% of all subscribers (network stations) behave as usual Internet users, mainly transmitting short packets (download requests) in the uplink direction. Accordingly, the behavior of the Internet users is represented by so-called uplink multimodal traffic model (Sec. 6.2.3). However, the average data rates between the Internet users is different and we define three uplink traffic classes, as presented in Tab. 6.4 [HrasLe03a].

The first uplink model has the lowest average data rate per user (0.75 kbps) and accordingly the largest mean interarrival time of the packets. We assume also that 40% of all stations in the network behave according to the traffic model M1. The average data rate is increased for two other uplink traffic models (2.5 and 7.5 kbps respectively for M2 and

Table 6.4 Traffic mix

	Model	Mean interarrival time of packets/s	Mean packet size/bytes	Average data rate/kbps	Share/%
M1	Uplink	3.55	332.5	0.75	40
M2	Uplink	1.06	332.5	2.5	20
M3	Uplink	0.35	332.5	7.5	10
M4	Downlink	0.88	822.33	7.5	10
M5	Downlink	0.26	822.33	25	10
M6	Downlink	0.07	822.33	100	10
	Average:	1.788	–	14.8	–

M3), whereas the interarrival time is decreased. Traffic models M2 and M3 are applied to 20 and 10% of all network stations respectively.

Each of the downlink traffic models (M4, M5, M6, Tab. 6.4) is applied to 10% of the stations with the average data rates per user of 7.5, 25, and 100 kbps. Note that the downlink traffic models are used to represent the users offering some Internet contents in the investigated PLC access network. The data produced by these traffic sources is transmitted in the uplink direction.

6.5.3 Simulation Results

The performance evaluation for the protocol comparison is carried out for the following three reservation MAC protocols:

- Extended ALOHA,
- Extended Active Polling, and
- Extended Hybrid-Two-step protocol.

All extended protocols implement piggybacking, dynamic backoff mechanism, and extended random access to the data channels for signaling purposes, as described in Sec. 6.3.2. The two-step protocol is implemented in its hybrid variant, ensuring random access to free request slots (Sec. 6.3.3). We observe two variants of the two-step protocol with different available number of pre-request slots; 5 and 40. The investigation is carried out by usage of the traffic mix, presented in Sec. 6.5.2, as a source model. All other model and simulation parameters (Sec. 6.2) are the same as in previous investigations (Sec. 6.3) using a simple retransmission mechanism for disturbed data packets (Sec. 6.4).

6.5.3.1 Network Utilization and Data Throughput

All three investigated protocols achieve the theoretical maximum network utilization, about 93% (Fig. 6.49). The remaining 7% of the network capacity is allocated for signaling (one of 15 channels) and it is never used for data transmission. Two-step and

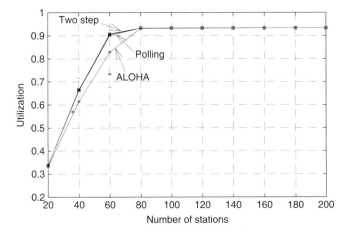

Figure 6.49 Average network utilization

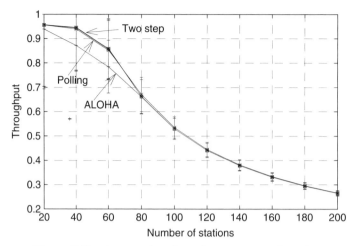

Figure 6.50 Average data throughput per network station

polling protocols show a nearly linear increase of the network utilization. However, polling achieves slightly lower utilization below so-called network saturation point (80 stations in the network) than both investigated variants of the two-step protocol (5 and 40 prerequest microslots per time slot, Sec. 6.5.1). On the other hand, ALOHA protocol behaves clearly worse than two-step and polling protocols below the saturation point.

As oppose to the behavior of network utilization, data throughput decreases with the increasing number of network stations (increasing network load, as presented in Fig. 6.50) and follows the results achieved for the network utilization, such as in the investigation of basic signaling protocols (Sec. 6.3.1). Below the network saturation point, the best behavior of the two-step protocol (both variants) can be again observed. Polling protocol achieves a slightly lower data throughput and ALOHA shows the worst behavior, as well.

6.5.3.2 Signaling Delay

In Fig. 6.51, it can be recognized that two-step protocols achieve the shortest signaling delay, even in the case that there are only five prerequest microslots. Polling protocol ensures shorter signaling delay than ALOHA almost in the entire investigated network load area. However, in the highly loaded network, the delay caused by ALOHA protocol is slightly shorter. This can be explained by application of piggybacking access method, which takes over most of the requests and releases the signaling channel. In this case, network stations, which are not able to use the piggybacking (because they are not active at the moment and their packet queue is empty), transmit the requests over the signaling channel. Since the signaling channel is rather released, the random access principles, such as ALOHA, ensure shorter signaling delay, as also shown in Sec. 6.3.2.

6.5.4 Provision of QoS in Two-step Reservation Protocol

In accordance with the simulation results presented in Sec. 6.5.3, we can conclude that the two-step protocol achieves the best performance among investigated reservation MAC

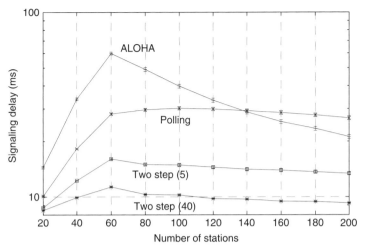

Figure 6.51 Mean signaling delay

protocols. As mentioned in Sec. 5.4.2, all reservation protocols allow an easy implementation of various mechanisms for traffic scheduling, due to the possibility of scheduling the transmission requests between the reservation procedure and the data transmission. However, in the two-step protocol, there is a further scheduling possibility ensured by the two-step procedure. Thus, it is possible to schedule the transmission prerequest before the stations are polled during the second protocol phase (Sec. 6.3.3). This is particularly important, if the distributed access control mechanism, combined with a signaling procedure with joint control messages, is applied (Sec. 6.1). In this case, there is no possibility of scheduling the transmission request if one-step reservation protocols are used; for example, ALOHA and polling-based solutions. On the other hand, the scheduling of the prerequests, which can be carried out in the two-step protocol ensures realization of different scheduling disciplines, such as realization of priorities, QoS control, and fairness.

The signaling delay achieved by the two-step protocol in the investigated network model remains below 20 ms for both its protocol variants; with 5 and 40 prerequest microslots within a signaling time slot (Fig. 6.51). This can be considered a reasonable signaling delay for data services, even ensuring realization of services with high time-critical requirements. Of course, the transmission time of the packets cannot be reduced only by application of an efficient MAC protocol. Therefore, for realization of data services with higher QoS requirements, it is necessary to implement an additional CAC mechanism (Sec. 5.4.3).

The transmission of voice can be implemented as a CBR service category, such as the classical telephony service, or as packet voice service, as is described in Sec. 4.4.2. In the first case, a transmission channel (e.g. OFDMA channel of 64 kbps) is allocated to a voice connection for its entire duration. The establishment of a voice connection is carried out in accordance with the signaling procedure, described in Sec. 6.1, where the signaling is used only for setting up the connection. Further signaling is only needed if the allocated channel is disturbed at the point at which a channel reallocation has to take place. So, with the signaling delay achieved in the investigated system (Fig. 6.51), it is possible to support the classical telephony service.

In the second case, network stations using the packet voice service transmit data that contains speech information only during so-called active periods of the talk (talkspurts). A network station using packet voice transmits a request at the beginning of a connection for its setup, such as in the case of the classical telephony service. After the connection is established, the network station has to send a request at the beginning of each talkspurt. If we assume that transmission channels for voice can be dynamically allocated (see Sec. 5.4.3 and Sec. 6.1.3), the voice station can start the transmission immediately or very shortly after the acknowledgment from the base station is received. In this way, the access delay can be reduced to its minimum, and the transmission delay of the voice packets consists mainly of the signaling delay. The delay limits for the voice service in access networks are set to relatively small values; for example, in wireless networks of 20 to 24 ms ([AlonAg00], [KoutPa01]), or of 25 ms to avoid the usage of echo cancelers [DaviBe96]. The maximum signaling delay in the investigated network model is below 20 ms (Fig. 6.51). So, in this case, the two-step protocol can fulfill the delay requirements.

6.6 Summary

To specify a reservation MAC protocol the following four functions have to be defined: reservation domain, signaling procedure, access control and signaling MAC protocol. An optimal reservation domain has to be chosen in accordance with transmitted telecommunications service. To avoid the transmission gaps occurring when the per-burst reservation is applied, the per-packet reservation domain is proposed for the realization of data transmission to improve network utilization. The signaling procedure and the access control have to be simple with a limited number of signaling messages, ensuring a low probability that the signaling exchange is affected by the disturbances. Among numerous proposals for signaling MAC protocols in different communications technologies, it is possible to identify two main protocol groups – protocols with random and with dedicated access.

The generic simulation model, used for the investigation of various signaling MAC protocols, implements the OFDMA/TDMA scheme, allowing implementation of multiple disturbance and traffic models. Two types of traffic models are considered – simple traffic models, representing the data traffic causing rare and frequent transmission requests, and multimodal traffic models, representing a nearly realistic behavior of Internet users. Two disturbance models are applied to allow investigations of lightly and heavily disturbed PLC networks.

Signaling delay, evaluated in the network using ALOHA protocol, is significantly shorter than in the network with polling in the case of rare transmission requests. In the case of frequent transmission requests, ALOHA protocol collapses and polling has significantly better performance. The protocol performance can be improved by the application of various protocol extensions. So, application of extended random access, using free data channels for signaling, improves network performance significantly in the low network load area as well as the piggybacking access method in the high loaded networks. On the other hand, with application of dynamic backoff mechanism, protocols with random access can be stabilized. Generally, it can be concluded that polling protocols, implemented in their advanced variants, have some advantages, and as opposed to advanced ALOHA protocols, they always achieve the theoretical maximum network utilization. Furthermore, the polling-based reservation protocols can be improved by the application of the active polling access method, reducing the signaling delay in high network load area.

A further reduction of signaling delays in the medium network load area is only possible, if the number of active stations is decreased, which can be ensured by the division of the polling procedure into two phases, building a so-called two-step reservation protocol – those are a prepolling phase, used for estimation of active network stations, and a polling phase, including the standard polling procedure of the active stations. The two-step protocol displays better performances than all other investigated one-step protocol solutions. Despite the more complex two-step signaling procedure compared with one-step protocols, the two-step protocol is not disadvantageous and it is robust against disturbances. To improve the performance of PLC networks operating under unfavorable noise conditions, an ARQ-plus mechanism without shifting is proposed to be applied in both one-step and two-step reservation MAC protocols using per-packet reservation principle.

Appendix A

A.1 Abbreviations

ACK	Acknowledgement
ADSL	Asymmetrical Digital Subscriber Line
ARQ	Automatic Repeat reQuest
ASK	Amplitude Shift Keying
ATM	Asynchronous Transfer Mode
AWGN	Additive White Gaussian Noise
BCH	Bose-Chaudhuri-Hocqunghem
BER	Bit Error Rate
BPSK	Binary Phase Shift Keying
BS	Base Station (main station for PLC or mobile wireless networks)
CAC	Connection Admission Control
CATV	Cable TV
CDM	Code Division Multiplex
CDMA	Code Division Multiple Access
CENELEC	Comité Europeen de Normalisation Electrotechnique
CFS	Carrier Frequency Systems
CISPR	Comité International Spécial des Perturbations Radio-electrique
CL	Controlled Load
CP	Cyclic Prefix
CPRMA	Centralized PRMA
CRC	Cyclic Redundancy Check
CRP	Collision Resolution Protocol
CSMA	Carrier Sense Multiple Access
CSMA/CA	CSMA with Collision Avoidance
CSMA/CD	CSMA with Collision Detection
DAB	Digital Audio Broadcasting
DAMA	Demand Assignment Multiple Access
DECT	Digital Enhanced Cordless Telecommunications Standard

DPRMA	Dynamic PRMA
DQDB	Distributed Queue Dual Bus
DQRAP	Distributed Queueing Random Access Protocol
DS-CDMA	Direct Sequence CDMA
DSL	Digital Subscriber Line
DSSS	Direct-Sequence Spread-Spectrum
EM	Electromagnetic
EMC	Electromagnetic Compatibility
EME	Electromagnetic Emission
EMI	Electromagnetic Interference
EMS	Electromagnetic Susceptibility
EIB	European Installation BUS
ETSI	European Telecommunications Standards Institute
EY-NPMA	Elimination Yield-Non-Preemptive Priority Multiple Access
FCC	Federal Communications Commission
FDD	Frequency Division Duplex
FDDI	Fiber Distributed Data Interface
FDMA	Frequency Division Multiple Access
FEC	Forward Error Correction
FFT	Fast Fourier Transform
FH-CDMA	Frequency Hopping CDMA
FHSS	Frequency Hopping Spread Spectrum
FSK	Frequency Shift Keying
FTP	File Transfer Protocol
GPRS	General Packet Radio Service
GS	Guaranteed Service
GSM	Global System for Mobile Communications
HF	High Frequency
HFC	Hybrid Fiber Coax
HTML	Hyper Text Markup Language
IAT	Interarrival Time
ICI	Inter-Channel Interference
IEC	International Electrotechnical Commission
IEEE	Institute of Electrical and Electronics Engineers
IDFT	Inverse Discrete Fourier Transform
IFFT	Inverse Fast Fourier Transform
IP	Internet Protocol
ISAP	Identifier Splitting Algorithm Combined with Polling
ISDN	Integrated Services Digital Network
ISI	Inter-Symbol Interference
ISMA	Inhibit Sense Multiple Access
ISMA/CA	ISMA with Collision Avoidance
ISMA/CD	ISMA with Collision Detection
ISO	International Standardization Organization
ITE	Information Terminal Equipment
LAN	Local Area Network

LCL	Longitudinal Conversion Loss
LEO	Low Earth Orbit
LFSR	Linear Feedback Shift Register
LLC	Logical Link Control
MAC	Medium Access Control
MAI	Multiple Access Interference
MC-CDMA	Multi-carrier CDMA
MC-DS-CDMA	Multi-carrier DS-CDMA
MCM	Multi-carrier Modulation
MCSS	Multi-carrier Spread-Spectrum
MDMA	Minimum-Delay Multi-Access
MEO	Medium Earth Orbit
M-PSK	M-ary Phase shift Keying
M-QAM	M-ary Quadrature Amplitude Modulation
MSAP	Mini-Slotted Alternating Priorities
MT	Mobile Terminal
MT-CDMA	Multi-tone CDMA
MV PLC	Medium-voltage PLC
NAK	Negative Acknowledgement
NRC	Non-Recursive Convolutional
OFDM	Orthogonal Frequency Division Multiplexing
OFDMA	OFDM Access
OSI	Open Systems Interconnection
PAN-SIM	PLC Access Network Simulator
PDF	Probability Distribution Function
pdf	Probability Density Function
PER	Packet Error Ratio
PLC	PowerLine Communications
PN	Pseudo-Noise
PNS	Pseudo-Noise Sequence
PODA	Priority-Oriented Demand Assignment
PRMA	Packet Reservation Multiple Access
psd	Power Spectral Density
QAM	Quadrature Amplitude Modulation
QoS	Quality of Service
QPSK	Quadrature Phase Shift keying
RA/DAMA	Random Access DAMA
RCS	Ripple Carrier Signaling
RSC	Recursive Systematic Convolutional
RTT	Round-Trip Time
SF	Spreading Factor
SNR	Signal to Noise Ratio
SS	Spread-spectrum
SSMA	Spread-spectrum Multiple Access
SSRG	Simple Shift Register Generator
TCL	Transversal Conversion Loss

TCP	Transmission Control Protocol
TDD	Time Division Duplex
TDMA	Time Division Multiple Access
TH-CDMA	Time Hopping CDMA
UMTS	Universal Mobile Telecommunications System
USB	Universal Serial Bus
VoIP	Voice over IP
WATM	Wireless ATM
WGN	White Gaussian Noise
WLAN	Wireless LAN
WLL	Wireless Local Loop
WWW	World Wide Web
YATS	Yet Another Tiny Simulator

References

[AcamKr] A. S. Acampora, S. Krishnamurthy, M. Zorzi, *Media Access Protocols for Use with Smart Array Antennas to Enable Wireless Multimedia Applications*, Center for Wireless Communications, University of California, San Diego, CA, USA.

[AkyiLe99] I. F. Akyilidz, D. A. Levine, I. Joe, A slotted CDMA protocol with BER scheduling for wireless multimedia networks, *IEEE/ACM Transactions on Networking*, **7**(2), 146–158 April 1999.

[AkyiMc99] I. F. Akyildiz, J. McNair, L. C. Martorell, R. Puigjaner, Y. Yesha, Medium access control protocols for multimedia traffic in wireless networks, *IEEE Network*, 39–47 July–August 1999.

[AlonAg00] L. Alonso, R. Agusti, O. Sallent, A near-optimum MAC protocol based on the distributed queueing random access protocol (DQRAP) for a CDMA mobile communication system, *IEEE Journal on Selected Areas in Communications*, **18**(9), 1701–1718 September 2000.

[Andr99] C. Andren, *A Comparison of Frequency Hopping and Direct Sequence Spread Spectrum Modulation for IEEE 802.11 Applications at 2.4 GHz*, Published in Electronic Engineering Times, December 1999; online available on http://www.eetasia.com/ARTICLES/1999DEC/1999DEC15_ICD_RFD_AN.PDF.

[AndrMa03] G. T. Andreou, E. K. Manitsas, D. P. Labridis, P. L. katsis, F. N. Pavilidou, P. S. Dokopoulos, Finite element characterisation of LV power distribution lines for high frequency communications signals, *Proceedings of the 7th International Symposium on Power-Line Communications and its Applications (ISPLC)*, Kyoto, Japan, 109–119 March 26–28, 2003.

[Arli] M. Arlitt, *Characterizing Web User Sessions*, Hewlett-Packard Laboratories, Palo Alto, CA, USA.

[BahaSa99] A. R. S. Bahai, B. R. Salzberg, *Multi-carrier Digital Communications: Theory and Applications of OFDM*, Kluwer Academic/Plenum Publishers, New York, 1992.

[BanwGa01] T. C. Banwell, S. Galli, A new approach to the modelling of the transfer function of the power line channel, *Proceedings of the 5th International Symposium on Power-Line Communications and its Applications (ISPLC)*, Malmö, Sweden, 319–324 April 4–6, 2001.

[Baum03] M. Baumann, *YATS – Yet Another Tiny Simulator, User's and Programmer's Manual, Version 0.4*, Dresden University of Technology, Chair for Telecommunications, Dresden, Germany, 2003, http://www.ifn.et.tu-dresden.de/TK/.

[BeardFr01] C. C. Beard, V. S. Frost, Prioritized resource allocation for stressed networks, *IEE/ACM Transactions on Networking*, **9**(5), October 2001.

[BegaBo00] K. Begain, G. Bolch, H. Herold, *Practical Performance Modelling, Application of MOSEL Language*, Kluwer Academic Publishers, 2000.

[BegaEr00] K. Begain, M. Ermel, A. Haidine, H. Hrasnica, M. Stantcheva, R. Lehnert, Modeling of a PLC network, *First Polish-German Teletraffic Symposium (PGTS2000)*, Dresden, Germany, September 24–26, 2000.

[Beny03] D. Benyoucef, A new statistical model of the noise power density spectrum for powerline communications, *Proceedings of the 7th International Symposium on Power-Line Communications and its Applications (ISPLC)*, Kyoto, Japan, 136–141 March 26–28, 2003.

[BerrGl93] C. Berro, A. glavieux, P. Thitimajshima, Near shannon limit error correcting coding and decoding: turbo codes, *Proceedings of IEEE International Conference on Communication (ICC '93)*, Geneva, May 1993.

[Bing00] B. Bing, Stabilization of the randomized slotted ALOHA protocol without the use of channel feedback information, *IEEE Communications Letters*, **4**(8), 249–251 August 2000.

[Bing02] M. Bingeman, *Symbol-Based Turbo Codes for Wireless Communications*, MSc Thesis, Electrical and Computer Engineering Department, University of Waterloo, Canada, 2002.

[Bore00] M. S. Borella, *Source Models of Network Game Traffic*, 3Com Corp, USA, 2000.

[Bumi03] G. Bumiller, System architecture for power-line communication and consequences for modulation and multiple access, *7th International Symposium on Power-Line Communications and its Applications (ISPLC2003)*, Kyoto, Japan, March 26–28, 2003.

[BumiPi03] G. Bumiller, N. Pirschel, Airfield ground lighting automation system realised with power-line communication, *7th International Symposium on Power-Line Communications and its Applications (ISPLC2003)*, Kyoto, Japan, March 26–28, 2003.

[CameZu00] F. Cameron, M. Zukerman, M. Ivanovic, S. Saravanabavananthan, R. Hewawasam, A deadlock model for a multi-service medium access control protocol employing multi-slot N-ary stack algorithm (msSTART), *Wireless Networks*, **6**, 391–399 2000.

[Chan00] W. C. Chan, *Performance Analysis of Telecommunications and Local Area Networks*, Kluwer Academic Publishers, Boston, Dordrecht, London, 2000, ISBN 0-7923-7701-X.

[ChlaFa97] I. Chlamtac, A. Farago, H. Zhang, Time-spread multiple-access (TSMA) protocols for multihop mobile radio networks, *IEEE/ACM Transactions on Networking*, **5**(6), 804–812 December 1997.

[ChoiSh96] S. Choi, K. G. Shin, Centralized wireless MAC protocols using slotted aloha and dynamic TDD transmission, *Performance Evaluation*, **27–28**, 331–346 1996.

[Cimi85] L. Cimini, Analysis and simulation of a digital mobile channel using orthogonal frequency division multiplexing, *IEEE Transaction Communications*, **COM-33**(7), 665–675, July 1985.

[ConnRyu99] D. P. Connors, B. Ryu, S. Dao, Modeling and simulation of broadband satellite networks; part I: medium access control for QoS Provisioning, *IEEE Communications Magazine*, 72–79 March 1999.

[Croz99] S. Crozier, Turbo-code design issues: trellis termination methods, interleaving strategies, and implementation complexity, *Invited Presentation for Panel Session on "Application of Turbo Codes" at International Conference on Communications (ICC'99)*, Vancouver, British Columbia, Canada, June 6–10, 1999.

[CrozLo99] S. Crozier, J. Lodge, P. Guinand, A. Hunt, Performance of turbo codes with relative prime and golden interleaving strategies, *Sixth international Mobile Satellite Conference (IMSC'99)*, Ottawa, Canada, June 16–18, 1999.

[DaviBe96] K. David, T. Benkner, *Digitale Mobilfunksysteme*, B.G. Teubner, Stuttgart, Germany, 1996, ISBN 3-519-06181-3, in German.

[DelFa01] E. Del Re, R. Fantacci, S. Morosi, R. Seravalle, Comparison of CDMA and OFDM systems for broadband downstream communications on low voltage power grid, *5th International Symposium on Power-Line Communications and its Applications (ISPLC2001)*, Malmö, Sweden, April 4–6, 2001.

[DengCh00] J. Deng, R. S. Chang, A nonpreemptive priority-based access control scheme for broadband ad hoc wireless atm local area networks, *IEEE Journal on Selected Areas in Communications*, **18**(9), September 2000.

[Dixi99] S. Dixit, Data rides high on high-speed remote access, *IEEE Communications Magazine*, 130–141 January 1999.

References

[Dost97] K. Dostert, *Telecommunications over the Power Distribution Grid – Possibilities and Limitations*, IIR-Powerline 6/97, Germany, 1997.

[Dost01] K. Dostert, *Powerline Communications*, Prentice Hall, 2001.

[Dost01a] K. Dostert, *Powerline-Kommunikation*, NET Zeitschrift für Kommunikationsmanagement, 2002, in German.

[DoufAr02] A. Doufexi, S. Armour, M. Butler, A. Nix, D. Bull, J. McGeehan, A comparison of the HIPERLAN/2 and IEEE 802.11a wireless LAN standards, *IEEE Communications Magazine*, 172–179 May 2002.

[EsmaKs02] T. Esmailian, F. R. Kschischang, P. G. Gulak, Capacity distribution of radiation-limited in-building power lines, *Proceedings of the 6th International Symposium on Power-Line Communications and its Applications (ISPLC)*, Athens, Greece, March 27–29, 2002.

[ETSI03] ETSI, *Power Line Telecommunications (PLT) Channel Characterization and Measurement Methods*, Technical Report ETSI TR 102 175 v1.1.1 (2003-03), European Telecommunications Standards Institute, 2003. Available online under www.etsi.org.

[FärbBo98] J. Färber, S. Bodamer, J. Charzinski, *Measurement and Modelling of Internet Traffic at Access Networks*, EUNICE Summer School, Munich, August 31–3 September, 1998.

[FazelPr99] K. Fazel, R. Prasad, Multi-carrier spread-spectrum, *ETT*, **10**(4), 347–350 July–August 1999.

[FeldGi] A. Feldmann, A. C. Gilbert, W. Willinger, T. G. Kurtz, The changing nature of network traffic: scaling phenomena, *ACM SIGCOMM*, Computer Communication Review

[FentBr01] D. Fenton, P. Brown, Some aspects of benchmarking high frequency radiated emissions from wireline communication systems in the near and far fields, *Proceedings of the 5th International Symposium on Power-Line Communications and its Applications (ISPLC)*, Malmö, Sweden, April 4–6, 2001.

[FerrCa03] M. Ferreiro, M. Cacheda, C. Mosquera, A low complexity all-digital ds-ss transceiver for powerline communications, *Proceedings of the 7th International Symposium on Power-Line Communications and its Applications (ISPLC)*, Kyoto, Japan, March 26–28, 2003.

[FleuKo02] B. H. Fleury, A. Kocian, *Spread Spectrum Technique and its Application to DS/CDMA*, Center for PersonalKommunikation, Aalborg University, Denmark, 2002, Online available on http://cpk.auc.dk/dicom/E02/core.pdf

[Fort91] P. J. Fortier, *Handbook of LAN Technology*, Intertex Publications, McGraw-Hill, Inc, New York, USA, 1991.

[FrigLe01] J. F. Frigon, V. C. M. Leung, A pseudo-Bayesian ALOHA algorithm with mixed priorities, *Wireless Networks*, 7, 55–63 2001.

[FrigLe01a] J. F. Frigon, V. C. M. Leung, H. C. B. Chan, Dynamic reservation TDMA protocol for wireless ATM networks, *IEEE Journal on Selected Areas in Communications*, **19**(2), 370–383 February 2001.

[GanzPh01] A. Ganz, A. Phonphoem, Robust superpoll chaining protocol for IEEE 802.11 wireless LANs in support of multimedia applications, *Wireless Networks*, 7, 65–73 2001.

[GargSn96] V. K. Garg, E. L. Sneed, Digital wireless local loop system, *IEEE Communications Magazine*, 112–115 October 1996

[Garo03] V. Garousi, *Methods to Reduce Memory Requirements of Turbo Codes*, Dissertation in Electrical and Computer Engineering, University of Waterloo, Canada, 2003.

[Goed95] J. Goedbloed, *Electromagnetic Compatibility*, Prentice Hall, New York, 1995.

[HaidHr02] A. Haidine, H. Hrasnica, R. Lehnert, MAC protocols for powerline communications: ALOHA, polling and their derivatives, *ITG-Fachtagung "Neue Kommunikationsanwendungen in modernen Netzen"*, Duisburg, Germany, 28 February–1 March, 2002.

[Hans00] D. Hansen, Megabits per second on 50 Hz Power Lines? *Proceedings of the International Wroclaw Symposium on Electromagnetic Compatibility EMC 2000*, Wroclaw, Poland, June 27–30, 2000.

[HaraPr97] S. Hara, R. Prasad, Overview of multicarrier CDMA, *IEEE Communications Magazine*, 126–133 December 1997.

[Hens02] C. Hensen, CISPR 22 Compliance test of power-line transmission systems, *Proceedings of the 6th International Symposium on Power-Line Communications and its Applications (ISPLC)*, Athens, Greece, March 27–29, 2002.

[Hern97] E. J. Hernandez-Valencia, Architectures for broadband residential IP services over CATV networks, *IEEE Network*, January/February 1997.

[Hooi98] O. G. Hooijen, On the channel capacity of the residential power circuit used as a digital communications medium, *IEEE Communications Letters*, **2**(10), October 1998.

[HomePlug] http://www.homeplug.org/

[HoudtBl00] B. Van Houdt, C. Blondia, Analysis of an identifier splitting algorithm combined with polling (ISAP) for contention resolution in a wireless access network, *IEEE Journal on Selected Areas in Communications*, **18**(11), 2345–2355 November 2000.

[Hras03] H. Hrasnica, *Medium Access Control Protocols for Powerline Communications Networks*, Dissertation, Dresden University of Technology, Dresden, Germany, 2003.

[HrasHa00] H. Hrasnica, A. Haidine, Modeling MAC layer for powerline communications networks, *Internet, Performance and Control of Network Systems, Part of SPIE's Symposium on Information Technologies*, Boston, MA, USA, November 5–8, 2000.

[HrasHa01] H. Hrasnica, A. Haidine, R. Lehnert, Performance comparison of reservation MAC protocols for broadband powerline communications networks, *SPIE's International Symposium ITCom2001: The Convergence of Information Technologies and Communications – Conference: "Internet, Performance and QoS"*, Denver CO, USA, August 19–24, 2001.

[HrasHa01a] H. Hrasnica, A. Haidine, R. Lehnert, Reservation MAC protocols for powerline communications, *5th International Symposium on Power-Line Communications and its Applications (ISPLC2001)*, Malmö, Sweden, April 4–6, 2001.

[HrasHa01b] H. Hrasnica, A. Haidine, R. Lehnert, *Powerline Communications im Anschlussbereich*, VDE Verlag, Germany, 48–53 NTZ 7-8/2001, in German.

[HrasLe00] H. Hrasnica, R. Lehnert, Powerline communications in telecommunication access area (Powerline Communications im TK-Zugangsbereich), *VDE World Microtechnologies Congress (MICRO.tec2000)*, ETG-Fachtagung und Forum: Verteilungsnetze im liberalisierten Markt, Expo 2000, Hannover, Germany, September 25–27, 2000.

[HrasLe00a] H. Hrasnica, R. Lehnert, Powerline communications for access networks – performance study of the MAC layer, *III International Conference on Telecommunications (BIHTEL 2000) "Telecommunication Networks"*, Sarajevo, Bosnia, Herzegovina, October 23–25, 2000.

[HrasLe01] H. Hrasnica, R. Lehnert, Simulation study of ALOHA based reservation MAC protocols for broadband PLC access networks, *XVIII Symposium on Information and Communication Technologies*, Sarajevo, Bosnia, Herzegovina, November 26–28, 2001.

[HrasLe02] H. Hrasnica, R. Lehnert, Extended ALOHA and hybrid polling reservation MAC protocols for broadband powerline communications networks, *XVIII World Telecommunications Congress (WTC2002)*, Paris, France, September 22–27, 2002.

[HrasLe02a] H. Hrasnica, R. Lehnert, Performance analysis of polling based reservation MAC protocols for broadband PLC access networks, *XIVth International Symposium on Services and Local Access (ISSLS2002)*, Seoul, Korea, April 14–18, 2002.

[HrasLe02b] H. Hrasnica, R. Lehnert, Performance analysis of two-step reservation MAC protocols for broadband PLC access networks, *6th International Symposium on Power-Line Communications and its Applications (ISPLC2002)*, Athens, Greece, March 27–29, 2002.

[HrasLe02c] H. Hrasnica, R. Lehnert, Performance analysis of error handling methods applied to a broadband PLC access network, *SPIE's International Symposium ITCom2002: Showcasing Communication, Networking, Computing, and Storage Technologies and Applications – Conference: "Internet, Performance and QoS"*, Boston, MA, USA, 29 July–1 August, 2002.

[HrasLe03] H. Hrasnica, R. Lehnert, Investigation of MAC protocols for broadband PLC networks under realistic traffic conditions, *7th International Symposium on Power-Line Communications and its Applications (ISPLC2003)*, Kyoto, Japan, March 26–28, 2003.

References

[HrasLe03a] H. Hrasnica, R. Lehnert, Performance comparison of reservation MAC protocols for PLC access networks under specific traffic conditions, *SPIE's International Symposium ITCom2003: Information Technologies and Communications – Conference: "Internet, Performance and QoS"*, Orlando, Florida, USA, September 7–11, 2003.

[HubbSa97] Y. C. Hubbel, L. M. Sanders, A comparison of the IRIDIUM and AMPS systems, *IEEE Network*, March/April 1997.

[Iano02] M. Ianoz, *Electromagnetic Effects Due to PLC and Work Progress in Different Standardization Bodies*, Final Report, EMC Group – Swiss Federal Institute of Technology, November 2002.

[IEC89] IEC, *Electromagnetic Compatibility, International Electrotechnical Vocabulary*, Chapter 161, IEC Publication 50(161), Geneva, January 1989.

[IEC01] IEC, *Electromagnetic Compatibility: The Role and Contribution of IEC Standards*, International Electrotechnical Commission, Lists of EMC Publications in IEC, Updated Version of February 2001.

[ieee90] Institute of Electrical and Electronics Engineers, IEEE Standard 802.6: Overview and Architecture, IEEE Std. 802, 1990.

[Ims99] L. A. Ims, Wireline broadband access networks, *Telektronikk*, 2/3, 73–87 1999.

[itu-t93] ISDN Service Capabilities, ITU-T Recommendation I.210, 03/93.

[JancWo00] T. Janczak, J. Wozniak, Performance analysis of HIPERLAN type 1, *First Polish-German Teletraffic Symposium (PGTS2000)*, Dresden, Germany, September 24–26, 2000.

[Joe00] I. Joe, A novel adaptive hybrid ARQ scheme for wireless ATM networks, *Wireless Networks*, 6, 211–219 2000.

[John90] M. J. Johnson, *Proceedings of the IFIP WG 6.1/WG 6.4 second international workshop on protocols for high-speed networks*, Palo Alto, CA, USA, ISBN 0 444 88932 9, November 27–29, 1990.

[JudgTa00] G. Judge, F. Takawira, Spread-spectrum CDMA packet radio MAC protocol using channel overload detection and blocking, *Wireless Networks*, 6, 467–479 2000.

[JungWa98] V. Jung, H. J. Warnecke, *Handbuch für die Telekommunikation*, Springer Verlag Berlin, Heidelberg, 1998, ISBN 3-540-62631-X -, in German.

[Kade91] F. Kaderali, *Digitale Kommunikationstechnik I*, Friedr. Vieweg & Sohn Verlagsgesellschaft mbH, Braunschweig, Germany, 1991, ISBN 3 528 04710 0, in German.

[KaldMe00] R. Kalden, I. Meirick, M. Meyer, Wireless Internet access based on GPRS, *IEEE Personal Communications*, 8–18 April 2000.

[KellWa99] R. Keller, B. Walke, G. Fettweis, G. Bostelmann, K. H. Möhrmann, C. Herrmann, R. Kraemer, Wireless ATM for broadband multimedia wireless access: the ATM mobil Project, *IEEE Personal Communications*, 66–80 October 1999

[Klein75] L. Kleinrock, *Queueing Systems, Volume I: Theory*, John Wiley & Sons, Inc., USA, 1975, ISBN 0-471-49110-1.

[KoffRo02] I. Koffman, V. Roman, Broadband wireless access solutions based on OFDM access in IEEE 802.16, *IEEE Communications Magazine*, April 2002.

[KousEl99] M. A. Kousa, A. K. Elhakeem, H. Yang, Performance of ATM networks under hybrid ARQ/FEC error control scheme, *IEEE/ACM Transactions on Networking*, 7(6), 917–925 December 1999.

[KoutPa01] P. Koutsakis, M. Paterakis, Highly efficient voice-data integration over medium high capacity wireless TDMA channels, *Wireless Networks*, 7, 2001.

[KuriHa03] K. Kuri, Y. Hase, S. Ohmori, F. Takahashi, R. Kohno, Power channel coding and modulation considering frequency domain error characteristics, *Proceedings of the 7th International Symposium on Power-Line Communications and its Applications (ISPLC)*, Kyoto, Japan, March 26–28, 2003.

[LangSt00] T. Langguth, R. Steffen, M. Zeller, H. Steckenbiller, R. Knorr, Performance study of access control in power-line communication, *4th International Symposium on Power-Line Communications and its Applications (ISPLC2000)*, Limerick, Ireland, April 5–7, 2000.

[Lee00] L. H. C. Lee, *Error-Control Block Codes for Communications Engineers*, Artech House, February 2000

[LenzLu01] L. Lenzini, M. Luise, R. Reggiannini, CRDA: a collision resolution dynamic allocation MAC protocol to integrate data and voice in wireless networks, *IEEE Journal on Selected Areas in Communications*, **19**(6), 153–1163 June 2001.

[Li02] J. Li, *Low-Complexity, Capacity-Approaching Coding Schemes: Design, Analysis and Applications*, Dissertation in Electrical Engineering, Texas A&M University, Texas, December 2002.

[LinCo83] S. Lin, D. Costello, *Error Control Coding: Fundamentals and Applications*, Prentice Hall, Englewood Cliffs, NJ, 1983.

[Lind99] J. Lindner, MC-CDMA in the context of general multiuser/multisubchannel transmission methods, *ETT*, **10**(4), July–August 1999.

[LiuWu00] H. -H. Liu, J. -L. C. Wu, Packet telephony for the IEEE 802.11 wireless LAN, *IEEE Communications Letters*, **4**(9), 286–288 September 2000.

[LiSo01] Ye Li, N. R. Sollenberger, Clustered OFDM with channel estimation for high rate wireless data, *IEEE Transactions on Communications*, **49**(12), 2071–2076 December 2001.

[MaedaFe01] Y. Maeda, R. Feigel, A standardization plan for broadband access network transport, *IEEE Communications Magazine*, 166–172 July 2001.

[Mars03] I. Marshall, *Principles of Digital Communication: MAP Decoding of Convolutional Codes*, SYSC 5504 (ELG 6154), Fall 03/04, Department of Systems and Computer Engineering, Carleton University.

[MatsUm03] H. Matsuo, D. Umehara, M. Kawai, Y. Morihiro, An iterative detection for OFDM over impulsive noise channel, *Proceedings of the 6th International Symposium on Power-Line Communications and its Applications (ISPLC)*, Athens, Greece, March 27–29, 2002.

[Meel99a] J. Meel, *Spread Spectrum: Applications*, De Nayer Institute, Belgium, October 1999.

[Meel99b] J. Meel, *Spread Spectrum: Introduction*, De Nayer Instituut, Belgium, October 1999.

[Modi99] E. Modiano, An adaptive algorithm for optimizing the packet size used in wireless ARQ protocols, *Wireless Networks*, **5**, 279–286 1999.

[MoenBl01] M. Moeneclaey, M. Van Bladel, H. Sari, Sensitivity of multiple-access techniques to narrowband interference, *IEEE Transactions on Communications*, **49**(3), 497–505 March 2001.

[Moly97] J. W. Molyneux-Child, *EMC Shielding Materials – A designer's Guide*, Second Edition, Newnes, Oxford, 1997.

[Müller02] T. Müller, *Traffic Management Mechanisms for ATM-based Differentiated TCP/IP-Networks*, Dissertation, Dresden University of Technology, Dresden, Germany, 2002.

[NakaUm03] Y. Naka, D. Umehara, M. Kawai, Y. Morihiro, Veterbi decoding for convolutional code over class A noise channel, *Proceedings of the 7th International Symposium on Power-Line Communications and its Applications (ISPLC)*, Kyoto, Japan, March 26–28, 2003.

[NatkPa00] M. Natkaniec, A. R. Pach, An analysis of the modified Backoff mechanism for IEEE 802.11 networks, *First Polish-German Teletraffic Symposium (PGTS2000)*, Dresden, Germany, September 24–26, 2000.

[NB30] *Frequenzbereichszuweisungen an Funkdienste im Frequenzbereich von 9 kHz bis 30 MHz – NB30*, Regulierungsbehörde für Telekommunikation und Post – RegTP, Germany, 2002, http://www.regtp.de/, in German.

[NeePr00] R. van Nee, R. Prasad, *OFDM for Wireless Multimedia Communications*, Artech House Publishers, Boston, London, 2000, ISBN 0-89006-530-6.

[NewbYa03] J. Newbury, J. Yazdani, From narrow to broadband communications using the low voltage power distribution network, *Proceedings of the 7th International Symposium on Power-Line Communications and its Applications (ISPLC)*, Kyoto, Japan, March 26–28, 2003.

[NishSh03] T. Nishiyama, T. Shirai, M. Itami, K. Itoh, H. Aghvami, A study on controlling transmission power of carriers of OFDM signal combined with data symbol spreading using optimal data reconstruction, *Proceedings of the 7th International Symposium on Power-Line Communications and its Applications (ISPLC)*, Kyoto, Japan, March 26–28, 2003.

[NishNo02] T. Nishiyama, S. Nomura, M. Itami, K. Itoh, H. Aghvami, A study on controlling transmission power of carriers of OFDM signal combined with data symbol spreading in frequency domain, *Proceedings of the 6th International Symposium on Power-Line Communications and its Applications (ISPLC)*, Athens, Greece, March 27–29, 2002.

[NomuSh01] S. Nomura, T. Shirai, M. Itami, K. Itoh, A study on controlling transmission power of carriers of OFDM signal, *Proceedings of the 5th International Symposium on Power-Line Communications and its Applications (ISPLC)*, Malmö, Sweden, April 4–6, 2001.

[Onvu95] R. O. Onvural, *Asynchronous Transfer Mode Networks – Performance Issues*, Artech House, Inc., USA, 1995, ISBN 0-89006-804-6.

[OrthPo99] B. Orth, M. Pollakowski, *ADSL – zukunfträchtige Übertragungstechnologie*, 276–293, Deutsche Telekom Unterrichtsblätter; 5/1999, in German.

[palas00] Group of authors, PALAS deliverable No. 4, European PLC Regulatory Landscape; June 2000.

[palas01a] Group of authors, PALAS deliverable No. 3 Technical Simulation Tool, February 2001.

[Peyr99] H. Peyravi, Medium access control protocols performance in satellite communications, *IEEE Communications Magazine*, 62–71 March 1999.

[Phil00] H. Philipps, Development of a statistical model for powerline communications channels, *Proceedings of the 4th International Symposium on Power-Line Communications and its Applications (ISPLC)*, Limerick, Ireland, April 5–7, 2000.

[PLCforum] http://www.plcforum.com/

[Pras98] R. Prasad, *Universal Wireless Personal Communications*, Artech House, Boston, London, 1998, ISBN 0-89006-958-1.

[Pris96] F. D. Priscoli, Smooth migration from gsm system to UMTS for multimedia services, *Wireless Networks*, **2**, 1996.

[Proa95] J. G. Proakis, *Digital Communications*, Third Edition, McGraw-Hill, New York, 1995.

[QiuCh00] X. Qiu, K. Chawla, L. F. Chang, J. Chuang, N. Sollenberger, J. Whitehead, RLC/MAC design alternatives for supporting integrated services over EGPRS, *IEEE Personal Communications*, 20–33 April 2000.

[RA96] Radiocommunications Agency, *Report on the Spectrum Audit of the Band – 9 kHz to 28000 kHz*, Spectrum management Section, October 1996. Online available at http://www.radio.gov.uk/publication/ra_info/ra305/ra305.htm.

[Rayc99] D. Raychaudhuri, Wireless ATM networks: technology status and future directions, *Proceedings of the IEEE*, **87**(10), October 1999.

[RegTP] Regulierungsbehörde für Telekommunikation und Post (RegTP); www.regtp.de.

[ReyesGo99] A. Reyes-Lecuona, E. Gonzalez-Parada, E. Casilari, J. C. Casasola, A. Diaz-Estrella, A page-oriented WWW traffic model for wireless system simulations, *16th International Teletraffic Congress (ITC16)*, Edinburgh, UK, 1999.

[Rodr02] M. R. D. Rodrigues, *Modelling and Performance Assessment of OFDM Communication Systems in the Presence of Non-linearities*, PhD Thesis, Department of Electronic and Electrical Engineering, University College, London, October 2002.

[RohlMa99] H. Rohling, T. May, K. Bruninghaus, R. Grunheid, Broad-band OFDM radio transmission for multimedia applications, *Proceedings of the IEEE*, **87**, 1778–1789, October 1999.

[RomSi90] R. Rom, M. Sidi, *Multiple Access Protocols – Performance and Analysis*, Springer-Verlag, New York, 1990.

[SahiTe99] Z. Sahinpglu, S. Tekinay, On multimedia networks: self-similar traffic and network performance, *IEEE Communications Magazine*, January 1999.

[SchnBr99] M. Schnell, I. De Broeck, U. Sorger, A promising new wideband multiple-access scheme for future mobile communications systems, *ETT*, **10**(4), 417–425 July–August, 1999.

[Schu99] C. Schuler, *Design and Implementation of an Adaptive Error Control Protocol*, GMD-Forschungszentrum Informationstechnik Gmbh, GMD Research Series, N°21/1999.

[SchuSc00] W. Schulz, S. Schwarze, Comparison of CDMA and OFDM for data communications on the medium voltage power grid, *International Symposium on Power-Line Communications and its Applications (ISPLC2000)*, Limerick, Ireland, April 5–7, 2000.

[Shan49] C. E. Shanon, *The Mathematical Theory of Communication*, Illinois Press, 1949.

[SharAl01] O. Sharon, E. Altman, An efficient polling MAC for wireless LANs, *IEEE/ACM Transactions on Networking*, **9**(4), 439–451 August 2001.

[ShirNo02] T. Shirai, S. Nomura, M. Itami, K. Itoh, Study on reduction of the affectation of impulse noise in OFDM transmission, *Proceedings of the 6th International Symposium on Power-Line Communications and its Applications (ISPLC)*, Athens, Greece, March 27–29, 2002.

[SiwkoRu01] J. Siwko, I. Rubin, Connection admission control for capacity-varying networks with stochastic change times, *IEEE/ACM Transactions on Networking*, **9**(3), 351–359 June 2001.

[Somm02] D. Sommer, *Beiträge zur Anwendung codierter OFDM-Modulation für drahtlose Übertragungssysteme*, Dissertation, Fakultät Elektrotechnik und Informationstechnik der Technischen Universität Dresden, Deutschland, April 2002, in German.

[SteeHa99] R. Steele, L. Hanzo, *Mobile Radio Communications: Second and Third Generation Cellular and WATM Systems*, Second Edition, Wiley, New York, 1999.

[Stev94] R. Stevens, *TCP/IP Illustrated Vol. 1 – The Protocols*, Addison-Wesley, 1994.

[Stro01] P. Strong, Regulatory & consumer acceptance of powerline products, *Proceedings of the 5th International Symposium on Power-Line Communications and its Applications (ISPLC)*, Malmö, Sweden, April 4–6, 2001.

[StroOt02] E. Stroem, T. Ottosson, A. Svensson, *An Introduction to Spread Spectrum Systems*, Department of Signals and Systems, Chalmers University of technology, Goeteborg, Sweden, 2002.

[TachNa02] S. Tachikawa, M. Nari, M. Hamamura, Power line data transmission using OFDM and DS/SS systems, *6th International Symposium on Power-Line Communications and its Applications (ISPLC2002)*, Athens, Greece, March 27–29, 2002.

[Tane98] A. S. Tanenbaum, *Computer Networks*, Third edition, Prentice-Hall Inc., USA, 1998, German edition.

[TayCh01] Y. C. Tay, K. C. Chua, A capacity analysis for the IEEE 802.11 MAC protocol, *Wireless Networks*, **7**, 159–171 2001.

[Tiha95] L. Tihanyi, *Electromagnetic Compatibility in Power Electronics*, IEEE Press, The Institute of Electrical and Electronics Engineers, New York, 1995.

[TrabCh] C. Trabelsi, H. K. Choi, Simulation and modeling of MAC protocols for wireless packet data networks using web traffic, *2nd IS-95 UMTS Workshop*.

[TranSt01] P. Tran-Gia, D. Staehle, K. Leibnitz, Source traffic modeling of wireless applications, *AEÜ International Journal of Electronics and Communications*, (1), 2001

[UmehKa02] D. Umehara, M. Kawai, Y. Morihiro, An iterative detection for M-ary SS system over impulsive noise channel, *Proceedings of the 6th International Symposium on Power-Line Communications and its Applications (ISPLC)*, Athens, Greece, March 27–29, 2002.

[Vick00] R. Vick, Radiated emission of domestic main wiring caused by power-line communications systems, *Proceedings of the International Wroclaw Symposium on Electromagnetic Compatibility EMC 2000*, Wroclaw, Poland, June 27–30, 2000.

[Vite98] A. J. Viterbi, *CDMA – Principles of Spread Spectrum Communication*, Addison Wesley Longman, Inc., USA, 1998, ISBN 0-201-63374-4.

[Walke99] B. H. Walke, *Mobile Radio Networks – Networking and Protocols*, John Wiley & Sons Ltd, Chichester, UK, 1999, ISBN 0-471-97595-8.

[Wong02] T. F. Wong, *Spread Spectrum and Code Division Multiple Access: Introduction to Spread spectrum Communications*, Electrical and Computer Engineering, University of Florida, Fall 2002; available online on http://www.wireless.ece.ufl.edu/eel6503/

[WongCh99] C. Y. Wong, R. S. Cheng, K. B. Letaief, R. D. Murch, Multiuser OFDM with adaptive subcarrier, bit and power allocation, *IEEE Journal on Selected Areas in Communications*, **17**(10), 1747–1758 October 1999.

References

[Zimm00]   M. Zimmermann, *Energieverteilnetze als Zugangsmedium für Telekommunikationsdienste*, Dissertation, Shaker Verlag, Aachen, Germany, 2000, ISBN 3-8265-7664-0, ISSN 0945-0823, in German.

[ZimmDo00]   M. Zimmermann, K. Dostert, An analysis of the broadband noise scenario in powerline networks, *International Symposium on Powerline Communications and its Applications (ISPLC2000)*, Limerick, Ireland, April 5–7, 2000.

[ZimmDo00a]   M. Zimmermann, K. Dostert, The low voltage distribution network as last mile access network – signal propagation and noise scenario in the HF- range, *AEÜ International Journal of Electronics and Communications*, (1), 13–22 2000.

[ZimmDo02]   M. Zimmermann, K. Dostert, A multipath model for the powerline channel, *IEEE Transactions on Communications*, **50**(4), 553–557 April 2002.

[ZhuCo01]   C. Zhu, M. S. Corson, A five-phase reservation protocol (FPRP) for mobile ad hoc networks, *Wireless Networks*, **7**, 371–384 2001.

[Yang98]   S. C. Yang, *CDMA RF System Engineering*, Artech House, Inc., USA, 1998, ISBN 0-89006-991-3.

Index

ABR, 123
Access delay, 224, 227
Access network, 8
Active polling, 175, 236, 241
Admission policy, 189
ALOHA, 153, 154, 219, 226
Allocation message, 198
Arbitration protocol, 154, 169
ARQ, 97, 111, 246
ARQ-plus, 247
Autocorrelation, 141, 143, 147

Background noise, 70
Backlog, 159
Base station, 41, 49
Bearer Service, 114
Best effort service, 123
BHC code, 103
Block code, 99
Blocking probability, 187, 207
Broadband PLC, 19

CAC mechanism, 189, 190
CBR, 123
CDMA, 128, 135
Channel allocation, 189
Channel availability, 207
Circuit switched, 200, 211
Coding, 87

Collision, 154, 156, 219
Collision avoidance, 167
Collision elimination, 168
Collision resolution, 157
Collision resolving, 159
Collision probability, 157
Conducted emission, 58, 61, 69
Connection level reservation, 196
Convolution code, 104
Contention protocol, 154
Contention window, 158, 186, 231
Control message, 198
Controlled load service, 122
Coupling factor, 63
CRC code, 103
Cross-correlation, 140, 147
Cross-product, 141
CSMA, 153
CSMA/CA, 167
CSMA/CD, 167
CSMA protocol, 160
Cyclic prefix, 84
Cyclic code, 102

Data segmentation, 130
Data throughput, 155, 228
Dedicated access, 204, 218
Digital subscriber line, 12
Direct sequence spread spectrum, 91, 95

Distribution network, 27
Distributed coordination function, 178
Disturbance, 34, 70, 75
Downlink/downstream, 50, 181
Dropping probability, 187
DS-CDMA, 136
Duplex mode, 181
Dynamic access, 153
Dynamic backoff mechanism, 158, 230, 241
Dynamic duplex mode, 184

Electromagnetic Compatibility, 33, 55
Electromagnetic emission, 56
Electromagnetic interference, 57
Electromagnetic susceptibility, 56
Error handling, 97, 244
Extended active polling, 251
Extended ALOHA, 251
Extended random access, 232, 241

Fairness, 188
Fast re-signaling, 245
Fast Fourier Transform, 86
Frequency division duplex, 181
FDMA, 12, 132
Fixed asymmetric mode, 184
Fixed access, 153, 196
Forward error correction (FEC), 97, 98
FH-CDMA, 136
Frequency hopping spread spectrum, 92
Frequency reuse factor, 149

Go-back-N ARQ, 112, 246
Gold code, 146
Guaranteed service, 122
Guard time, 83

Hamming code, 102
Hamming distance, 101
Hamming weight, 101
Hard Blocking, 150
Hidden terminals, 166
Holding time, 116
Hybrid MAC protocol, 175, 180
Hybrid two-step protocol, 250

Impulsive noise, 71, 73
In-Home PLC, 21, 47
Interarrival time, 73, 116
Inter-Carrier Interference, 83
Interleaving, 87, 108
Inter-Symbol Interference, 83
Inverse Fast Fourier Transform, 86
ISMA protocol, 167
ISO/OSI reference model, 79

Linear Feedback Shift Register, 144
Link metric, 148
Loading factor, 149
Logical channel, 199
Loss probability, 187

m-sequence, 144
MAC layer, 36, 81, 125, 205
MAC protocol, 153
Mapping, 87
MC-CDMA, 140
Modulation, 82
Multi Carrier Modulation, 82, 140
Multimodal traffic models, 213
Multipath channel, 53
Multiple Access Interference, 150
Multiple access scheme, 125, 128

Narrowband PLC, 16
Network topology, 39
Network section, 40
Network segmentation, 43
Network utilization, 155, 164, 206, 220, 232
Noise, 70
Nonpersistent CSMA, 160

OFDM, 82
OFDM access, 133
OFDMA/TDMA, 134
OFDM/TDMA, 129
Optimal retransmission probability, 158

Packet delay, 224
Packet switched channel, 200, 201
PAN-SIM, 215

Partial correlation, 141
Per-burst reservation, 196
Per-packet reservation, 197
Persistent CSMA, 160
Phase Shift Keying, 87
Piggybacking, 230, 241
PLC access network, 19, 40
PLC Gateway, 25, 46
PLC Repeater, 24, 46
Polling, 172
Point coordination function, 180
Prepolling, 240
Probability distribution function, 116
Probability density function, 116
Propagation delay, 171, 174
Protected two-step protocol, 245
Pseudo-Bayesian algorithm, 159
Pseudo-noise sequence generator, 91, 143
Pseudorandom Sequence, 143

Quadrature Amplitude Modulation, 87
Quality of service, 118
QoS control, 187
QoS guarantee, 187
QoS parameter, 205

Radiated emission, 58, 61, 67
Random access, 204, 218
Reallocation message, 199
Reed-Solomon code, 103
Reservation domain, 196
Reservation protocol, 176, 195, 198, 218
Resource sharing, 125, 151
Reverse sequence generation, 145
Rivest's Pseudo-Bayesian algorithm, 159
Round-trip time, 171, 173, 223
RTS/CTS, 178

Send-and-Wait ARQ, 111
Selective-reject ARQ, 113, 246
Service classification, 121

Shanon's capacity, 98
Signaling delay, 224, 234, 254
Signaling message, 198, 201
Simple shift register generator, 144
Simulation model, 208, 215, 217
Slotted ALOHA, 156, 223
Soft Blocking, 150
Splitting algorithm, 159
Spreading factor, 90
Spreading gain, 90
Spread-Spectrum modulation, 89
Supply network, 14, 39
Symmetric duplex mode, 184

TDMA, 128
TH-CDMA, 138
Time division duplex, 181
Token passing, 169
Token-Ring, 169
Token-bus, 170
Traffic class, 122
Traffic control, 181
Traffic model, 119, 211, 213
Transfer function, 53
Transfer time, 171
Transmission channel, 52
Transmission cable, 53
Transmission delay, 224, 227
Transmission request, 198, 218
Tree topology, 46
Turbo code, 107
Two-step reservation protocol, 240, 242

Uplink/Upstream, 50, 181
UBR, 123
User modeling, 210

VBR, 123
Voice activity factor, 149

Wireless local access network, 11
Wireless local loop, 10